Atomically Heterogeneous Systems

Sudeep Bhattacharjee

Atomically Heterogeneous Systems

Low Energy Ion Beams from Plasmas for Material Modification

Springer

Sudeep Bhattacharjee
Department of Physics
Indian Institute of Technology Kanpur
Kanpur, Uttar Pradesh, India

ISBN 978-981-95-9732-1 ISBN 978-981-95-9733-8 (eBook)
https://doi.org/10.1007/978-981-95-9733-8

This Springer imprint is published by the registered company Springer Nature Singapore Pte Ltd.
The registered company address is: 152 Beach Road, #21-01/04 Gateway East, Singapore 189721, Singapore

To

whom I owe the most in this journey

Jayeeta for enriching it in immeasurable ways

Sayak and Syon, for making it meaningful and full of joy

Students working in the laboratory both past and present, who are the real warriors and provided wings to my ideas

Foreword

The book *Atomically Heterogeneous Systems—Low Energy Ion Beams from Plasmas for Material Modification* will serve as a textbook for advanced graduate and post-graduate students enrolled in the physics course. Professor Sudeep Bhattacharjee is a renowned academician working extensively in the areas of plasma science and technology, as well as the applications in the areas of modification of materials. I am impressed to see that he has taken time out to write this unique book of current interest very systematically for the benefit of the students and researchers.

The book covers both theoretical and experimental treatments. The fundamental and advanced topics of plasmas and ion beams, plasma ion sources and ion beam extraction from plasmas are dealt with utmost care. While discussing the fundamentals of plasmas and ion beams in the introductory chapters, the author demonstrates these aspects using advanced research results wherever necessary.

Different types of available ion sources in the world are discussed including the plasma ion source developed by the author. The book discusses the interaction of low-energy ion beams derived from plasmas, and their utilization for engineered materials with novel properties. The book further discusses the applications of these engineered materials in optical, electronic, space and field emission devices. Various other topics covered in this book are the generation of broad and focused low-energy micro-ion beamlets and their control, creation of atomically heterogeneous systems including case studies of research on modifying electrical, optical including field emission properties of materials.

The detailed coverage and pedagogical tools make this an ideal textbook for students enrolled in physics course and wanting to learn this exciting subject, including those desiring to enter the research area of utilizing low energy ion beams for material modification, and their novel applications. It will also be an excellent reference book for researchers and professionals in the areas of material science and engineering. The book is expected to be of immense benefit to industrial scientists and personnel who are planning to work on large scale modification of material properties for applications.

This is the first book of its kind to cover the creation of novel properties in materials using subsurface excitations by low energy plasma ion beams (0–5 keV) in details.

The book brings together the physics and applications of low energy ion beams with exhaustive coverage of plasma source from where the ions are derived to the new market driven applications, without compromising on the underlying fundamental physics bringing about the novel changes. The book also deals comprehensively with broad and multi-disciplinary research topics covering the areas of plasma physics, ion beam physics, materials science and their properties including optical, electrical and field emission.

Dinakar Kanjilal
Inter-University Accelerator Center (IUAC)
New Delhi, India

Preface

This book is a result of research work carried out by the author, his students, and collaborators since 1996, on the generation of compact microwave plasmas in multicusp magnetic fields and thereafter utilizing the plasmas for low-energy (0–3 keV) ion beam extraction, and modification of material properties for novel applications.

For extraction of ion beams, compact plasmas are helpful as it can facilitate production of ion beams with sizes comparable or smaller than the plasma dimension. Since microwaves are employed for plasma generation, a compact plasma chamber size ensures that a higher wave power density can be maintained, leading to high-density plasma production and subsequent generation of higher ion beam currents. In this regard, the magnetic multicusp was found to be a useful plasma confinement device, as it provides excellent radial and axial plasma confinement. The axial confinement being brought about by end plugging the multicusp, which is realized by having magnets of opposite polarity. Moreover, inside the multicusp there is a large region near the center where the magnetic field is very small (null at the center) which facilitates ion beam extraction through the central unmagnetized plasma regions. Additionally, it was earlier found that a multicusp magnetic field facilitates plasma production in a waveguide (which is usually the cylindrical plasma chamber) with a dimension below the geometrical cutoff and favors sustenance of overdense plasmas (plasmas with a density higher than the wave cutoff density). The ion beams are extracted from the compact plasma by employing electrostatic fields. Usually a plasma electrode made up of a conducting material such as stainless steel, having a tiny aperture through which the beam is extracted, is placed at one end of the multicusp, and another electrode (called the collector) is kept a few cm away to which a negative high voltage is applied for extracting the ion beams from the plasma. The material sample to be processed with ion beams is often placed on the collector. The distance between the collector and the plasma electrode is optimized so that beam divergence is minimal in order to ensure beam uniformity on the sample surface.

In this book, research utilizing low-energy ion beams having energy in the range 0–3 keV is described. The low-energy ions enter into the surface and subsurface layers of the material typically from a few to tens of nanometers inside the material. Since primarily inert gas ions such as Ar, Kr, He, and Ne are utilized, after these ions enter inside the substrate material they do not from any compound with the host material, rather an atomically heterogeneous system is formed with unique properties which did not exist in the parent material. This interesting aspect belongs to the subject matter of the book. In a few cases, ion beam from plasmas of molecular gases such as N_2 and H_2 has also been employed in some studies. The main materials which have been subjected to ion beam irradiation experiments include thin films of metals, such as Cu, Al, and Ag, having thickness of a few 100 nm.

The metallic thin films that have been modified by ion beams are found to have unique optical, electrical (sheet resistivity), and field emission properties. Such ion beam-modified metallic thin films can be utilized for a variety of emerging applications, for example, in optical mirrors in spacecrafts and other optical sensors in space vehicles, making optical filters, in microelectronics by tuning the electrical sheet resistance, in field emission devices for generation of bright electron sources for use in electron guns, or high-resolution electron microscopy and in vacuum electronic devices for microwave generation. It is also found to be of great relevance to the study of discharge of induced charges from spacecrafts. When ion beams impinge upon the substrate of different materials such as Al and Cu, they lead to micro-structural formations on the substrate, and such structures have relevance to the studies on the effect of ion beams on plasma-facing components in fusion devices, for example, in divertors in Tokamak plasmas.

The book starts with an introduction to plasmas and discusses about the plasma state of matter in Chap. 1. The fundamental plasma properties are detailed including the criterion for a collection of charge particles to be called as a plasma. The various methods of plasma creation and confinement in the laboratory are discussed. The principal physical processes and the governing equations to describe a plasma are outlined. Toward the end of the chapter plasma sheath and generation of ion beam are discussed.

In Chap. 2, different types of plasma ion sources and the methodology of beam extraction from the plasmas are presented. A variety of existing plasma-based ion sources, their advantages and limitations have been discussed, including the two most commonly used plasma sources, namely, the RF and the microwave plasma ion sources. A special mention has been made of the popular microwave plasma-based ECR ion sources which are employed in accelerators around the world for the generation of highly charged ion beams. Toward the end of this chapter, the microwave plasma-based multicusp ion source that has been found to provide stable high-density plasmas and ion beams, and that have been employed in the research works presented in this book is discussed.

Ion beams can be of different types. In Chap. 3, the generation of broad and focused multi-ion beamlets which are utilized depending upon the specific application of the ion beams is discussed. The experimental setup is described including the plasma electrode apertures for the generation of the two types of ion beams. AXCEL-INP and SIMION simulations help to corroborate the experimental results of beam extraction utilizing the different plasma electrode apertures. These two simulation tools are the most commonly used in ion beam optics design in the literature. The chapter also discusses about the space charge effects during ion beam extraction and a technique for ion beamlet control in the case of focused multi-ion beamlets. The current density for a circular aperture is derived, and the total beam current is measured for the different plasma electrode apertures. The beam uniformity for the focused multi-ion beamlets is presented.

The focused micro-ion beamlets facilitate surface patterning and localized modification. However, a methodology needed to be developed in order to manipulate the micro-ion beamlets. Chapter 4 is an elaboration of the research work on generation of focused micro-ion beamlets, and a methodology to control the beamlets is discussed for surface patterning. The design of the beamlet switching circuit is presented, and, in particular, the stopping potential for the ion beamlet is determined, a potential which completely blocks beam transmission. The generation of surface patterns on Al foil which are localized resistive regions is demonstrated. The spot size of individual beamlets and micron-scale resistivity measurements are carried out.

In Chap. 5, the creation, characterization, and applications of atomically heterogeneous systems are discussed. Starting from the definition, it is shown how such systems can be created in the laboratory. Thereafter, the diagnostic techniques for the characterization of such systems are described such as optical and electrical. Finally, a summary of the experiments carried out with the atomically heterogeneous system is provided. This chapter provides the much-needed introduction to the subject matter discussed in this book.

Since the ion beams are derived from gaseous plasmas confined in magnetic multicusps, it is important that the characteristics of the plasmas, particularly the plasma parameters and their spatial profiles, are investigated for different types of multicusp, so that a particular type of multicusp which provides a high density can be chosen for future experiments. Chapter 6 therefore presents some basic experiments on plasma and ion beams. In this chapter, the magnetic confinement of the plasma using three different types of multicusps such as the octupole (8 poles), decapole (10 poles), and duo-decapole (12 poles) is discussed. Measurements of the magnetic field inside the multicusp including the radial profiles of the plasma density and electron temperature are presented. The number of ions impinging on the substrate per unit area and per unit time is provided by the ion beam flux. The flux measurements are carried out as a function of microwave power, discharge pressure, and for plasmas of different gases such as Ne, Ar, He, and Kr. The ion saturation current and beam uniformity are measured at the end of the multicusp. The chapter ends with a discussion on the sample preparation techniques.

Chapters 7–10 are on different case studies related to applications of atomically heterogeneous systems. In particular, Chaps. 7 and 8 are on measurements of electrical sheet resistance both at room and at low temperatures. Chapter 9 to parts of Chap. 10 are on optical studies where different optical constants such as refractive index, skin depth, and optical conductivity, including reflectivity, transmissivity, and absorptivity of metallic thin films, are measured as a function of the ion beam fluence and energies, and for different ionic species and substrate materials. Chapter 10 also deals with field emission studies, where due to ion irradiation micro-conical structures are created on the metallic surfaces, this coupled with the effect of a reduced work function due to embedded subsurface atomic impurities gives rise to enhanced electron field emission current. We present the different types of case studies carried out in each chapter, in a more detailed manner below.

To begin with, Chap. 7 provides a theoretical background of the origin of resistivity in metals. The electron transport is discussed in single and polycrystalline films including the temperature dependence of resistivity in a metal. Meticulous experiments are performed in metallic thin films to measure the electrical sheet resistance as a function of temperature. The experiments are performed at various fluences of argon ion beam at a fixed energy of 0.3 keV and in another set, the films are irradiated with different ionic species (H_2, N_2, Ne, Ar, Kr) (mass number 1–84) at a fixed fluence (9.6×10^{15} ions cm^{-2}).

In Chap. 8, the investigation reported in Chap. 7 is further elaborated. The temperature variation of sheet resistance in metals for pristine and ion beam-irradiated samples is carried out in greater details. The sheet resistivity and Debye temperatures in metallic nano-films of Ag, Cu, and Al are investigated as a function of ion beam-induced defects and impurities in a controlled manner. Ions of both atomic (Ne, Ar, and Kr) and molecular (H_2, N_2) gases are employed in the investigation and the number of defects and impurities in the nano-film are varied in a controlled manner by varying the ionic mass number (1–84) and beam fluence (8.7×10^{15}–1.4×10^{16} ions cm^{-2}). Low-temperature measurements for pristine and irradiated films are also performed to obtain the residual sheet resistance (RRS). The chapter provides an empirical formula relating the variation of RRS with beam fluence and ionic mass number and ends with a discussion on the diffusion of ions in matter and the available simulation tools.

Chapter 9 discusses the modification of optical properties of metallic thin films of copper, silver, gold, and aluminum, by Ar ion beams of 0.5 keV. Optical parameters such as the real and imaginary parts of the refractive index (n and k) are investigated as a function of ion beam fluence. The optical properties such as reflectivity, transmissivity, and absorptivity are investigated using UV-VIS-NIR spectrophotometry. The real and imaginary parts of the refractive index are measured by employing the pseudo-Brewster angle technique. The modification in the optical properties of the metallic thin films is discussed in light of theories that describe the optical properties of mixed media such as Maxwell-Garnett and the Bruggeman approximation.

In Chap. 10, the Kramers-Kronig (KK) technique for determining the optical constants of the ion beam-irradiated thin films of metals such as Ag, Al, and Cu at a fixed beam energy (0.5 keV) is discussed. The reflectivity of pristine MTF is measured using UV-VIS-NIR spectrophotometry having energy range of light from $\sim$1 to 4.96 eV (or wavelength of 250–1200 nm) and is used as an input parameter in the KK technique to determine the optical constants. The KK reflectivity transformation is employed to obtain the phase information, and further the optical constants. The Maxwell-Garnett (MG) approximation is used to determine the effective dielectric constants of mixed media, which are further corroborated using the pseudo-Brewster angle technique for four different laser wavelengths (405, 532, 632.8, and 670 nm).

The research is further continued but with a specific application in mind. In Chap. 10, the relevance of ion irradiation in a space environment and the impact of He ion beams of varying energies (0.5–3 keV) on space relevant materials such as Ag and Al are investigated. This chapter reports modifications in the reflectivity of Ag and Al metallic thin films upon irradiation of low-energy helium ion beams with varying energies (0.5, 1, 2, and 3 keV) and varying fluences ranging from 1.10×10^{16} cm^{-2} to 1.56×10^{16} cm^{-2}. The fluence of helium ions is chosen according to the 4 (1.1×10^{16} cm^{-2})-and 6 (1.56×10^{16} cm^{-2})-year doses of alpha particles encountered by a solar orbiter. The helium ions' dose corresponding to 4 and 6 years can be realized within a few seconds (~ 3 seconds) in the high flux that can be obtained from our plasma source.

Chapter 10 further discusses the modification in field emission properties when thin films of metals are irradiated with low-energy ion beams. The change in surface morphology leads to production of micro-conical protrusions which act as excellent electron field emitters. The basics of field emission are discussed including the underlying theories of micro-structural growths on the surface of metals. Field emission is influenced both by microstructures that are formed due to enhancement of the electric field, depending upon the aspect ratio of the structures, and due to formation of atomic heterogeneity, i.e., the implanted subsurface atoms, which lowers the work function (as was observed in the case of both hillocks and valley regions of the generated structures). X-ray photoelectron spectroscopy and Kelvin Probe Force Microscopy (KPFM) provided information on the extent of atomic heterogeneity and work function of the ion beam-modified surfaces.

A comparative field emission study from micro-conical structures created in Cu substrate by two different low-energy noble gas ion beams, i.e., by irradiating with a constant argon ion flux of 6.47×10^{15} cm^2s^{-1} and a krypton ion flux of 4.81×10^{15} cm^2s^{-1} at 2 keV energy is described in Chap. 10. For comparative purposes, the fluence of ion beams of both species was kept the same, and therefore the irradiation time was adjusted accordingly. The development of the structures (aspect ratio, areal distribution of hilly and valley regions) has been investigated as a function of ion beam fluence and the amount of foreign atomic impurities for each fluence (implanted Kr and Ar atoms). Dynamic scaling theory has been applied to understand the nature of the growth of surface morphology. The structural growths have also been looked at using KPFM and Transmission Electron Microscopy (TEM). The cross-sectional TEM results unveiled some interesting features such as the dimensions of the surface protrusions, including the formation of an amorphous and a damage layer. Chapter 11 summarizes the book and provides some future prospects of the variety of works carried out.

Kanpur, India Sudeep Bhattacharjee

Acknowledgments This book is a culmination of collaborative research works and support from many people whom I would like to gratefully acknowledge. First of all, I would like to acknowledge Prof. Hiroshi Amemiya, former Senior Scientist at the Institute of Physical and Chemical Research (RIKEN) and Guest Professor at Saitama University, Japan, who was my doctoral supervisor (1996–1999) and made me interested in microwave-generated plasmas. These electrodeless electromagnetic wave-generated plasmas offered a way to generate clean plasmas devoid of any contaminants. During the time the research was initiated, it was a challenging problem, to create an electromagnetic wave-generated plasma in a waveguide-type vessel having a cross-sectional dimension smaller than the geometrical cutoff value for wave propagation. The cutoff dimension therefore poses as a limit to wave propagation and therefore plasma generation and sustenance in such narrow cross-sectional waveguide-type plasma vessels, due to non-existence of propagating mode solutions of Maxwell's equations. I worked on methods to generate microwave plasmas in such narrow cross-sectional waveguides and found that utilization of a magnetic multicusp surrounding the waveguide helped to obviate the waveguide geometrical cutoff problem and offered an innovative way to create compact overdense plasmas, which could be utilized for the extraction of ion beams—an experimental investigation which became an important part of my doctoral thesis. The high-density compact plasma became suitable for the extraction of intense (high current) pencil-like ion beams. Such a device that utilizes a compact multicusp microwave plasma has been built in our laboratory at IIT Kanpur after I took up an academic position here and employed throughout the research works presented in this book, which forms the basis of low-energy ion beam source for interaction with materials.

I would like to acknowledge the works of my former Ph.D. students whose have worked under my guidance in the laboratory, for their doctoral degrees on interaction of low-energy ion beams with matter. I would like to acknowledge Dr. Abhishek Chowdhury for his contribution on the generation of broad and micro-focused ion beamlets. The creation and optimization of plasma electrodes with notched and through apertures including those created by laser drilling, and for the work on the modification of electrical properties by ion beams. Abhishek measured the temperature variation of sheet resistance for pristine and ion beam-irradiated samples and also implemented a technique for surface patterning using the focused micro-ion beamlets. The work on the determination of stopping potential, ion beamlet control, micro-resistivity, and current-voltage characteristics was carried out under my guidance by Dr. Chowdhury collaboratively with my other students Dr. Sanghamitro Chatterjee and Mr. Apurba Dutta, and I would like to acknowledge all of them. I am also happy to acknowledge the work of another former Ph.D. student Dr. Sanjeev Kumar Maurya who worked on focused ion beams.

Most of the research work on optical properties of low-energy ion beam-modified metallic thin films were carried out under my guidance by my former Ph.D. student Dr. Krishn Pal Singh whom I would like to acknowledge. Several experimental methods were undertaken by him to investigate the optical properties of ion beam-modified materials, such as spectrophotometry, pseudo-Brewster angle technique, and theoretical treatments such as the Kramers-Kronig technique, including the application of Maxwell-Garnett and Bruggeman approximation models, whose results were compared with the experimental results to understand the optical properties of mixed media. As an interesting application, the optical properties were studied in a space-like environment where the impact of He ion beams of varying energies (0.5–3 keV) and fluences was applied to mimic the 4 (1.1×10^{16} cm^{-2})-and 6 (1.56×10^{16} cm^{-2})-year doses of alpha particles encountered by a solar orbiter.

I would like to acknowledge the work on field emission from ion beam-modified metallic thin films and thin sheet of metals by my current Ph.D. student Ms. Jayashree Majumder. Her research work on creation of self-organized micro-conical arrays of protrusion in metals such as Cu and Al led to the finding that such micro-conical structures created by ion beam interaction with metals led to significant enhancement of field emission current. The existing Fowler-Nordheim relation was utilized to compare with results obtained from the experiments. It was found that due to ion beam irradiation there was a noticeable change in the work function of the materials, particularly in the valley and hilly regions of the ion beam-irradiated micro-conical metallic surfaces. A comparative study of field emission from micro-conical tips created by different low-energy noble gas ions has also been undertaken.

The variety of works carried out in this book required the use of multiple material characterization tools such as SEM, AFM, TEM, XPS, spectrophotometer, profilometer, thin film deposition systems, including closed cycle refrigerators for low-temperature measurements of electrical resistivity, and it was a good fortune to have most of these facilities at my academic institute—IIT Kanpur. I thank Prof. Ashish Garg for the AFM and KPFM measurements, the Centre of Nanosciences for the SEM measurements, Advanced Imaging for the TEM measurements, and the Advanced Center for Materials Science (ACMS) for the XPS measurements, all carried out at IIT Kanpur. I gratefully acknowledge Prof. A. K. Jha (in charge of Optics Laboratory), for providing equipment support from Optics Laboratory, IIT Kanpur to conduct some experiments. I also thank Prof. R. Vijaya for providing Profilometer and UV-VIS-NIR spectrophotometer facilities for thickness and reflectivity measurements. I thank Prof. Y. N. Mohapatra for allowing us to use the closed cycle refrigerator for low-temperature electrical measurements of ion beam-modified metallic thin films. I am grateful to Department of Materials Science and Engineering, IIT Kanpur for providing us with the SEM and EDX facilities which were used in some experiments. I am also grateful to DST Unit of Nanosciences and the Department of Chemical engineering, IIT Kanpur for use of AFM and laser micro-fabrication facility.

During research, we needed to fabricate several experimental components. I gratefully acknowledge the support of our Physics workshop at IIT Kanpur, where Mr. Omprakash and his colleagues fabricated these components which were of high technical quality and made in the least possible time.

Many of the works presented in this book are an outcome of research during the doctoral studies of afore-mentioned students, for whom I was the thesis supervisor.

We received valuable feedback during research from the peer-review committee members for the students, who are my faculty colleagues in the Department of Physics at IIT Kanpur. I would therefore like to acknowledge them. I thank Prof. M. K. Harbola, Prof. S. M. Tripathi (currently at IIT Delhi), and Prof. Aditya H. Kelkar for valuable suggestions on the work on optical investigations of ion beam-modified materials. I would like to thank Prof. Satyajit Banerjee and Prof. T. K. Ghosh for their valuable feedback on the field emission investigations of ion beam-modified materials. I would like to thank Prof. H. C. Verma (retired Professor of IIT Kanpur) and Prof. Anjan K. Gupta for fruitful discussions on the work on electrical measurements of ion beam-modified materials. Apart from them there are others who have helped us in many possible ways during the journey of research. I acknowledge fruitful discussions with Prof. John H. Booske, University of Wisconsin-Madison, United States, who was my former postdoctoral advisor, for fruitful discussions on the work on field emission from ion beam-modified materials. I thank Dr. Dipak Bhowmik who was my former Institute Postdoctoral Fellow working in our laboratory for helping the students with the Casa XPS software and in SRIM/TRIM simulations of ion beam interactions with materials. I also thank my current Ph.D. student Mr. Sushanta Barman for helping in the Python programming to evaluate the optical constants of the materials using KK technique and helping on improving the quality of a few figures in the book. I extend my heartfelt thanks to Dr. Joy Mukherjee who was my former Institute Postdoctoral Fellow working in our laboratory, who meticulously read several chapters of the book and improved the written sentences and style, including improving several figures in the book. I sincerely thank my other Ph.D. students Deepika, Ayesha Nanda, Swati Swagatika Mishra, and Sargam for helping out in various ways which helped me to complete this book.

There are a lot of costs involved in purchasing equipment, chemicals, and consumable items while carrying out experimental research. A significant cost is also involved in purchasing machine time related to the characterization facilities. I gratefully acknowledge our funding agencies, namely, Department of Science and Technology (DST)—SERB, Council of Scientific and Industrial Research (CSIR), and Indian Space Research Organization (ISRO), Government of India, for funding my research work, without the help of the funding agencies it would be impossible to carry out the experimental research works presented in the book. I also gratefully acknowledge the financial support received from our Institute and Physics Department in this regard.

Writing the book would simply not have become possible without the support received from people at Springer Nature. I would like to heartily thank Ms. Priya Vyas of Springer Nature for her support all along this journey, right from the time we started having discussions to conceptualize this book and deciding a suitable title, and for accommodating my deadline extension requests, which has been very helpful. I would also like to heartily thank Ms. Neha Sharma of Springer Nature who took care of the publication of the book from beginning of August 2024, and with her I had several discussions to improve the book and finally to bring the book to completion. Ms. Neha Sharma and Ms. Priya Vyas's help in answering my several queries, such as obtaining permission or on the front matter of the book, was immensely helpful, particularly toward the finishing stages of the book. I also would like to thankfully acknowledge the support from Ms. Rini Christy during the production stage of the book for providing me a checklist of items during submission and for all the guidance toward submission of the materials including evaluation of the chapter contents. I would like to acknowledge support from Mr. Ankit Upadhyay of IIT Kanpur for creating the figure files and arrangement of all the source files.

Finally, I would like to heartily acknowledge the support from my family members including my wife Jayeeta Bhattacharjee and sons Sayak Bhattacharjee and Syon Bhattacharjee for sacrificing their share of my time, during the entire duration of writing the book, and for the immense love and encouragement I received from them, which was undoubtedly instrumental in bringing the book to a completion. I would like to acknowledge the continuous encouragement received from my parents Mr. Bibhuti Bhusan Bhattacharjee and Mrs. Sipra Bhattacharjee, who have greatly motivated me and shared my joy in writing the book and for making me who I am. I would also like to acknowledge my parents-in-law, late Mr. Ranjit Kumar Ray and Mrs. Chandana Ray for their constant love, affection, and support toward my academic goals, the book is a small recognition of their support. I also thank my brothers Sandeep Bhattacharjee and Joydeep Bhattacharjee for their continuous love and support toward my academic endeavors.

Competing Interests The author has no competing interests to declare that are relevant to the content of this manuscript.

Contents

About the Author

Dr. Sudeep Bhattacharjee is currently Professor of highest academic grade (HAG) at the Department of Physics, Indian Institute of Technology, Kanpur. He obtained his Bachelor's and Master's degree in Physics from the University of North Bengal, and his Ph.D. degree from the Institute of Physical and Chemical Research (RIKEN), Japan in affiliation with Saitama University, Japan, under a Japanese Government Ministry of Education Science and Culture (Monbusho) Scholarship. His current major research areas are experimental plasma physics involving low temperature plasmas, focused ion beams for generation of nano and quantum beams, and low energy ion matter interactions for creation of atomically heterogeneous systems to bring about novel properties in materials. Dr. Bhattacharjee has published more than 100 refereed journal publications, and has written a book titled "Compact plasma and focused ion beams". Dr Bhattacharjee has received several awards such as the Buti Foundation Award (2009), India, the Australian Government Endeavour Research Award (2010), the Norwegian Government Specialist Exchange Scholarship (2006). At IIT Kanpur he received the Excellence in Teaching Award (2023) and held the Sanjay and Rachna Pradhan Chair Professorship (2017–2020). Dr. Bhattacharjee has been a member of the International scientific and organizing committee of prestigious International Conferences in Plasma Physics namely the International Conference on the Phenomena of Ionized Gases (ICPIG), the annual conference of the Association of Asia Pacific Physical Societies—Division of Plasma Physics (AAPPSDPP), and the International Congress on Plasma Physics (ICPP). Dr. Bhattacharjee has been a member of the Editorial Board of distinguished journals in Plasma Physics namely Review of Modern Plasma Physics (Springer), Plasma Research Express (IOP), and Frontiers in Physics, Plasma Physics (Frontiers).

Chapter 1
Introduction to Plasmas and Ion Beams

1.1 Plasma as the Fourth State of Matter

We know from our early school days, we learn that there are three states of matter, namely, solid, liquid, and gas. When a gas is heated to an extremely high temperature, such that the atoms comprising the gas are ionized forming positive ions, and electrons which have been knocked out from the neutral gas atoms during the process of ionization, and some unionized neutral particles—this is a state which represents the *fourth state of matter* and is known as a plasma. Welcome to the world of plasmas which are ubiquitous. They occur in nature in large-scale astrophysical objects such as the stars, the nebulae, and the supernovae. Our sun is an example of a very hot plasma. In the laboratory, plasmas are created in several types of devices from small-scale micro-plasmas to large volume plasmas which are confined in Tokamaks or Stellarators and investigated for fusion energy, to lessen our dependence on fossil fuels as a source of energy. Depending upon the method of producing a plasma the degree of ionization can be different in the plasma. We will discuss this in a later section. What sets the plasma state of matter apart is the collective behavior of its constituents, such as ions, electrons, and neutral gas particles which define a plasma.

Figure 1.1 shows the different varieties of plasmas that can be either created or formed in the laboratory, space, or outside (such as flames). Each type of plasma is characterized by the plasma density (plotted on the x-axis) and the electron temperature (plotted on the y-axis). One can see that plasmas can have a wide range of electron densities and temperatures. Figure 1.2, on the other hand, shows similar information with more plasma types and have been classified as relativistic, classical, strongly coupled, and quantum plasmas, depending upon the particle densities, temperatures, and relevant energies. In the laboratory, we mostly study low-pressure plasmas which constitute a majority of plasma-based ion sources. These days high-pressure plasmas created in the atmosphere are becoming quite popular due to their applications in the biomedical field. Both the low-pressure and atmospheric pressure plasmas belong to classical plasmas. There is another class of plasmas that are created at cryogenic

S. Bhattacharjee, *Atomically Heterogeneous Systems*,
https://doi.org/10.1007/978-981-95-9733-8_1

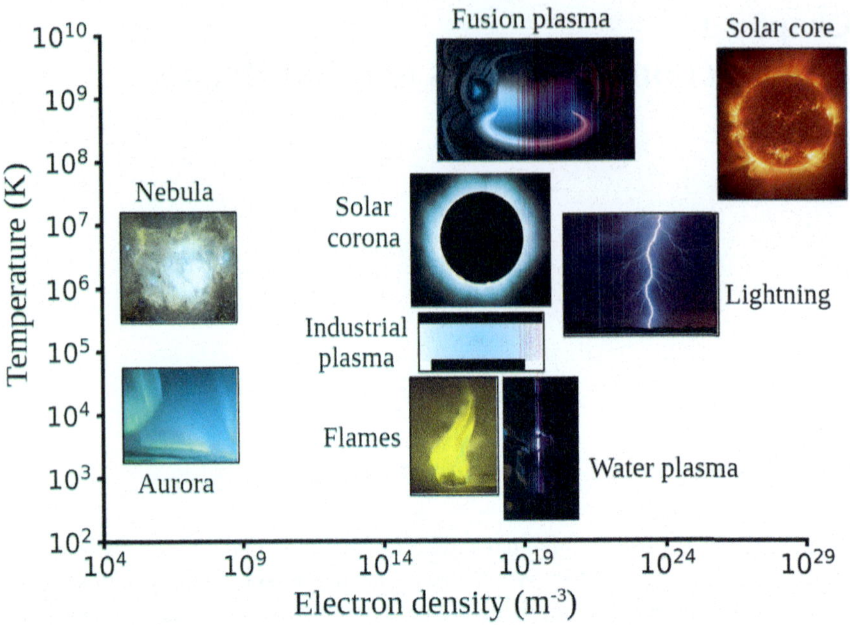

Fig. 1.1 A schematic diagram to show different types of plasmas depending upon the typical values of the plasma electron temperature (°K) and electron density (m^{-3})

temperatures [at liquid N_2 (80 K) or liquid He (4 K) temperatures], where the coupling parameter Γ, given by [1].

$$\Gamma = \frac{E_c}{\kappa_B T} = \frac{Q^2}{4\pi\varepsilon_o a\kappa_B T} \tag{1.1}$$

is >1, where E_c is the mean coulomb interaction energy of the particles and $\kappa_B T$ is the mean thermal energy, Q is the charge, T is the temperature, and $a = [3/(4\pi n)]^{1/3}$ the Wigner-Seitz radius (density n). When $\Gamma \gg 1$ the plasma is said to be strongly coupled. Note that in conventional low-pressure plasmas, kinetic energy of the particles dominates over the Coulomb interaction energy and the plasma is weakly coupled.

Under thermodynamic equilibrium, the degree of ionization and the electron temperature are closely related by Saha's equation [named after the Indian Scientist Meghnad Saha (1893–1956)], given by [2].

$$\frac{n_i}{n_n} = 3 \times 10^{27} T^{3/2} \frac{1}{n_i} e^{-U/T}, \tag{1.2}$$

where n_i is the plasma ion density in m^{-3}, n_n is the density of the neutral atoms in m^{-3}, T is the electron temperature in eV, and U is the ionization potential of the gas

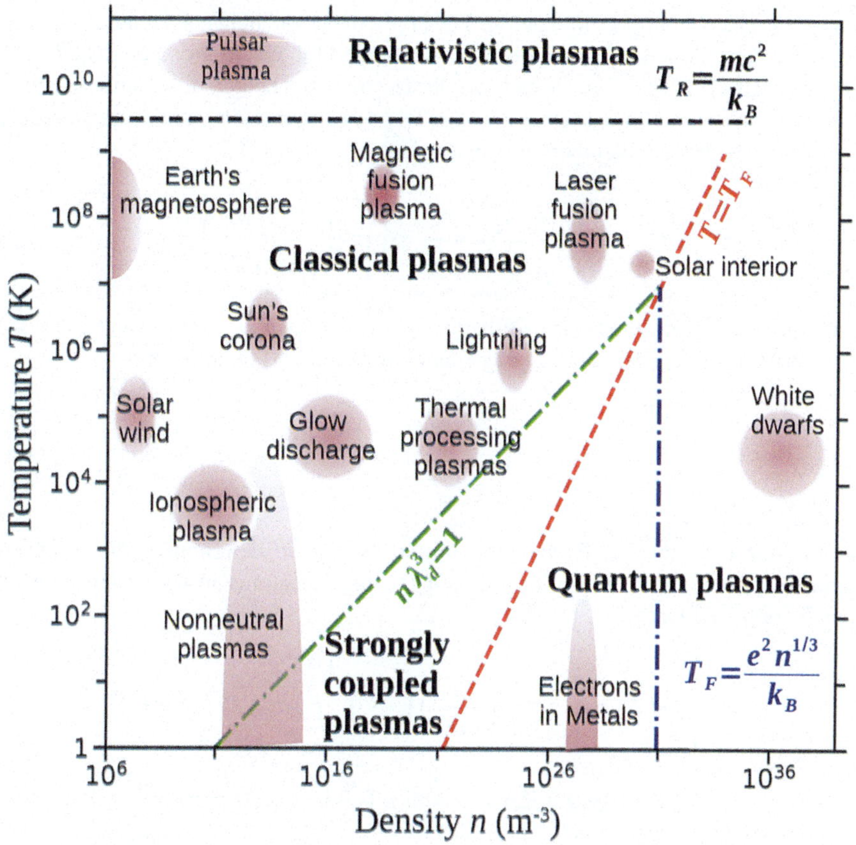

Fig. 1.2 A schematic diagram to show different types of plasmas depending upon the typical values of their separating the classical and quantum regimes.

in eV. By employing the above formula one can determine the degree of ionization or the ionization fraction, $\alpha = \frac{n_i}{n_i + n_n}$ of the plasma, and can broadly classify the plasma as strongly, moderately, or weakly ionized plasmas. For example, in a low-pressure plasma confined by a dipole magnet, α lies between 10^{-3} and 10^{-4} (0.1 and 0.01%) and in an atmospheric pressure plasma jet, α lies in the range 10^{-6}–10^{-7} (10^{-4}–10^{-5}%), which can be classified as a weakly ionized plasma. On the other hand, in a fusion plasma reactor such as in a Tokamak the plasma can be close to being fully ionized. Plasmas are utilized in many real-life applications such as in tube lights, plasma TV, semiconductor chips (IC), material processing, and material synthesis such as diamond and graphene and for plasma treatment in the medical field.

Plasma can be created in a variety of ways in the laboratory. Commonly used technique is electric field-induced ionization, where electrons are accelerated to high energies in the electric field of the energy source and can ionize the gas, a process known as electron impact ionization. The applied electric field can create a distribution of electron energies in the plasma often called the electron energy distribution function (EEDF). There are various standard normalized energy distributions such as Maxwellian (f_M), Druyvesteyn (f_D), and F4 (f_4) distributions, which are governed by the common formula mentioned below [3]:

$$f(E) = \frac{x\left[\Gamma\left(\frac{5}{2x}\right)\right]^{3/2}\sqrt{E}}{\left[\Gamma\left(\frac{5}{2x}\right)\right]^{3/2}\langle E\rangle^{3/2}}\exp\left[-\left(\frac{\Gamma\left(\frac{5}{2x}\right)E}{\Gamma\left(\frac{3}{2x}\right)\langle E\rangle}\right)\right], \tag{1.3}$$

where Γ is the complete Gamma function; $x = 1$ for f_M, $x = 2$ for f_D, and $x = 4$ for f_4 distribution; and $\langle E\rangle$ is the average energy of electrons obtained from

$$E = \int_{E\min}^{E\max} Ef(E)\mathrm{d}E. \tag{1.4}$$

Furthermore, the EEDF plays an important role in determining the electron-neutron collision and ionization rates of the neutral atoms in the plasma, which can be obtained from

$$\nu_{en/i} = N_g\langle\sigma_{en/i}\,\mathrm{v}\rangle = \left(\frac{p}{\kappa_B T}\right)\left(\frac{2e}{m_e}\right)\int_{E\min}^{E\max}\sigma_{en/i}E^{\frac{1}{2}}f(E)\mathrm{d}E, \tag{1.5}$$

where $N_g = p/\kappa_B T$ is the gas density with p being neutral pressure (in Pa), $\kappa_B = 1.38 \times 10^{-23}$ m^2 kg s^{-2} K^{-1}, $T = 300$ K, $e = 1.602 \times 10^{-19}$ C and $m_e = 9.109 \times 10^{-31}$ kg, κ_B being the Boltzmann constant, T the gas temperature taken as room temperature, e is the electronic charge and m_e is the mass of an electron, respectively, and $\mathrm{v} = \sqrt{2E/m_e}$ (in m/s) is the velocity of the electron in the electric field. The $\sigma_{en/i}$ (in m^2) is the cross section of the corresponding process which can be determined by two methods. Method-I involves formulae for respective processes, where the effective electron-neutron collision cross section is calculated using the following equation:

$$\sigma_{en} = \frac{E}{\langle E\rangle}Q^{\mathrm{m}}(E), \tag{1.6}$$

where $Q^{\mathrm{m}}(E)$ is the momentum transfer cross section of electrons for both inelastic and elastic collisions combined. For ionization cross section, the following formula is used:

$$\sigma_i = \pi\left(\frac{e}{4\pi\varepsilon_o E_i}\right)^2\left(\frac{E - E_i}{E_i}\right), \tag{1.7}$$

where $\varepsilon_o = 8.854 \times 10^{-12}$ F/m is the permittivity of free space and E_i is the ionization energy of the gas atoms. Method-II considers directly employing experimental data for $\sigma_{en/i}$ for electrons in the gas, for example, Argon gas.

1.2 Fundamental Plasma Properties

Plasma whose constituents are electrons, ions, and neutral particles show collective behavior which are guided by certain criteria. Therefore, any collection of charged particles cannot be called a plasma. The first criteria is known as *quasi-neutrality*, which means the number of electrons and ions in a plasma are approximately equal. This means one can take $n_i \cong n_e \cong n$, where n is a common density called the plasma density. This is important as the plasma as a whole remains electrically neutral. Before we discuss the other criteria for the state of matter to be called a plasma, it is important to talk about the characteristic length and time scales in a plasma, which are the Debye length λ_D and inverse of plasma frequency f_p, respectively. The angular plasma frequency is given by $\omega_p = (n_e e^2/m_e \epsilon_o)^{1/2}$. There is also another characteristic frequency known as the electron neutral collision frequency ν, which defines the rate at which electrons collide with neutral particles in a plasma, which could be neutral atoms or molecules. The mean free path for electron neutral collisions is given by $\lambda_m = 1/n_n\sigma$, where n_n is the neutral density (neutral atoms per m^3) and σ is the collision cross section. The mean time between collisions is given by $\tau = \lambda_m/\mathrm{v}$ and the mean frequency of collisions is $\tau^{-1} = \mathrm{v}/\lambda_m = n_n\sigma\mathrm{v}$. If we now average over all particles of all velocities v in a Maxwellian distribution, we have the collision frequency $\nu = n_n\langle\sigma\mathrm{v}\rangle$.

The concept of Debye length was introduced by Debye Huckel from his theory of electrolytes, where a positively charged ion in the ionic liquid is surrounded by electrons such that the potential due to the positive charge is shielded out by the electron cloud surrounding it. The distance at which the potential of the positive charge drops to $1/e$ of the value in the absence of the cloud is known as the Debye length. In an alike manner, electrons are also surrounded by ions. A similar phenomena happens in a plasma where an ion can be surrounded by electrons and vice versa, and thus Debye shielding is realized. The potential V at a distance r due to the shielding of a positively charged ion of charge Q by the electron cloud can be written as

$$V(r) = \frac{Q}{4\pi\varepsilon_o r} \mathrm{e}^{-\frac{r}{\lambda_D}}, \tag{1.8}$$

where ε_o is the permittivity of free space and λ_D is the Debye length given by $\lambda_D = (\varepsilon_o \kappa_B T_e/ne^2)^{1/2}$. The second criteria is that the *length of the plasma system* $L \gg \lambda_D$. In other words, there should be several Debye lengths in the system for it to adequately exhibit plasma properties. Therefore, any local potentials are shielded out within a distance of a few times λ_D. The third criteria for it to be called a plasma is that the *number of particles in the Debye sphere*, $N_D \gg 1$, for effective shielding to occur, that is:

$$n\left(\frac{4}{3}\pi\lambda_{\mathrm{D}}^{3}\right) \gg 1. \tag{1.9}$$

The plasma frequency mentioned above arises when a plasma is disturbed from its equilibrium position. Suppose we let an electric field fall instantaneously on a plasma which may be taken as a plasma slab, it is then seen that the electrons would be swayed to one side in the direction of the field, leaving behind the heavier ions, and as they move to one side, the ions would try to pull them back. During its reverse motion, the electrons would overshoot its equilibrium position due to inertia and this would set up an oscillation about the mean position with a frequency known as electron plasma frequency. The ions being much bulkier as compared to the electrons would not respond as quickly to the applied field. Another plasma criterion is that the *electron plasma frequency* $f_{\mathrm{p}} \gg \nu$, otherwise the oscillatory motion of the plasma would die out too soon due to the frequent electron neutral collisions.

1.3 Creation of Plasmas in the Laboratory

Plasmas can be created by a variety of ways such as the DC plasma, which occurs when a DC electric field is applied to a gas at low pressure. Ionization of the gas takes place to form a plasma. Plasmas can also be produced by applying an AC field, known as an AC plasma such as the one used in our Hg-containing fluorescent tube lights where AC frequency of 50 Hz and voltage of 220 V are applied. There are also plasmas which are created by electromagnetic waves in the radio frequency (RF) or microwave regime and are commonly known as an RF or microwave plasmas. Furthermore, plasmas can be created using lasers and they are known as laser-produced plasmas. For example, to create an oxygen plasma whose ionization potential is 13.6 eV requires the wavelength λ of incident photons <91 nm and lies in the far UV region. The wavelength of the laser needs to be chosen such that the energy of the photon exceeds the ionization threshold of the gas to be ionized to form a plasma. In the electromagnetic wave-assisted plasmas, the most commonly used RF frequency is 13.56 MHz and for this type of RF excitation two types of plasmas are well known, that is, the capacitively coupled (CCP) and the inductively coupled plasmas (ICP). The CCP is created between two parallel plates where one plate is the driven plate and the other plate is grounded. In the ICP, there is a coil that surrounds the quartz tube where the plasma is created. A special type of antenna known as the helicon wave antenna helps to generate helicon plasmas, which are known to support very high densities. There have also been reports of UHF and VHF plasma sources created by radio frequency waves in the 50–800 MHz regime and have also been considered favorable, as they provide commercially viable processing rates and substantial reduction in sheath voltages compared to RF plasmas. In the microwave regime, the 2.45 GHz frequency has been extensively employed for generation of microwave plasmas. Here, upon the application of a magnetic field, plasmas can be generated very efficiently by employing electron cyclotron resonance (ECR), a phenomenon which occurs when the wave frequency becomes equal to the cyclotron resonance frequency of the electrons in the magnetic field, and efficient transfer of power from the waves to the particles can be achieved. The ECR condition can be described by $\omega = \omega_c = eB/m$, where ω and ω_c are the angular wave and cyclotron frequency, respectively, and B is the magnetic field. In another example, where the waves are made to move along the surface of a dielectric in a plasma, surface wave plasmas can be realized. Recently atmospheric pressures plasmas have become very popular, and such plasmas have been created by employing high voltages usually in the DC or AC modes where pulsed or sinusoidal voltage waveforms are usually employed. In fusion plasmas, that are created in large devices such as Tokamaks, the plasma is initiated by running a current through a primary coil placed at the center of a Tokamak, which bring about a $\partial B/\partial t$ and thereby induces an electric field (through Maxwell's equation $\vec{\nabla} \times \vec{E} = -\frac{\partial \vec{B}}{\partial t}$) in the toroidal direction and brings about breakdown of the gas to form a plasma.

1.4 Plasma Confinement Methods

Plasmas that are created need to be confined else they would disappear due to processes such as recombination, where an ion and an electron recombine to form a neutral atom, and diffusion where the particle diffuses out of the system and gets lost. Diffusion can be governed by density and temperature gradients. There can also be plasma instablities which lead to loss of energy and particles from the plasma system. Therefore, plasmas need to be confined in a particular volume for carrying out experimental investigations, otherwise they would simply disappear. There are several types of confinement schemes such as by a combination of poloidal and toroidal magnetic fields which together produces a helical magnetic field that confines the plasma in a Tokamak. Other confinement schemes include the mirror machine, where two current-carrying coils in a Helmholtz-type configuration and spaced apart can create a uniform field in the central region and peaked fields at the location of the coils. Such a type of magnetic field configuration helps to confine the plasma, because the particles make mirror motion between region of higher magnetic fields and is confined. This type of confinement is widely employed in ECR ion sources around the world and will be discussed in a later chapter. Then there is the multicusp confinement where alternate polarity of either permanent magnets or current-carrying coils arranged in a circular manner creates a magnet null at the center, with the field increasing toward the wall of the multicusp. The plasma can very well be confined in the multicusp. Unlike the torus in a Tokamak, where the end losses are cancelled and the particles can go around the torus, in a multicusp the end losses can arise which is suppressed by end plugging of the multicusp, where the polarities of the magnets at both ends are reversed [4]. There is yet another interesting confinement scheme known as the dipole confinement, where a dipole magnetic field is created either by electric currents that are run through a levitated coil to produce a dipole field [5] or a single permanent is employed to create such a magnetic field [6]. It is found that plasmas can be very efficiently confined in such a dipole magnetic field, due to a balance of plasma pressure and the magnetic pressure.

1.5 Principal Physical Processes and Governing Equations to Describe a Plasma

There are many physical processes that go on inside a plasma. To begin with plasmas are generated by ionization which can happen through electron impact. Here an energetic electron interacts with the atoms in the gas to bring about ionization. The energy of the impacting electron should be at least equal to the ionization potential of the gas. In the ionization process, an ion and two electrons are produced. One electron is the slow striking electron and the other one is the electron ejected from the parent atom. There is also another type of ionization process called radiative ionization where a photon of sufficient energy is absorbed by an atom dissociating it into an ion and electron. Plasmas are lost by processes such as recombination and diffusion. Recombination is the reverse process of ionization where an ion can recombine with an electron to form a neutral atom. Recombination processes are of two types, namely, three-body recombination and radiative recombination. In three-body recombination, two electrons and an ion join to make a neutral atom and an electron. In the radiative recombination process, an electron and an ion combine into an atom, and a photon is emitted. There are two types of collision processes, namely, elastic and inelastic collisions. In the elastic collision process, an electron essentially bounces off an atom with the two particles retaining their identities as electron and an atom. In the inelastic process, there is exchange of energy which can lead to ionization and excitation. An excitation process leads to the formation of metastable atoms, whose energy is close to the ionization threshold; therefore, such metastable atoms can very easily be ionized even by a low-energy electron. As mentioned above, plasmas can be lost through a process of diffusion, the diffusion coefficient is given by $D = KT/mv$, where K is Boltzmann's constant, T is the temperature, m is the mass, and v is the collision frequency. The flux of each species of particles can be written as

$$\vec{\Gamma}_j = n\vec{\mathrm{v}}_j = \pm\mu_j n\vec{E} - D_j\vec{\nabla}n, \tag{1.10}$$

where n is the number density, $\mathrm{v_j}$ is the velocity of species j, μ_j is the mobility of species j, $\vec{E}$ is the electric field, and D_j is the diffusion coefficient of species j. The mobility is given by $=|q|D/\kappa_{\mathrm{B}}$, where the magnitude of the charge is given by $|q|$. The above equation shows that electrons and ions will have different fluxes which can lead to charge imbalance. If the plasma size $\gg$ than the Debye length, it must be quasi-neutral and different rates of diffusion of electrons and ions would somehow adjust so that the two species leave together. The electrons are lighter and have higher thermal velocities, they tend to leave the plasma first. The positive charge is left behind. However, an electric field is set up of such polarity as to retard the loss of electrons and accelerate the loss of ions, making both the species to diffuse with the same velocity. This is known as ambipolar diffusion. Setting $\vec{\Gamma}_e = \vec{\Gamma}_i$, the ambipolar electric field can be written as [7].

$$\vec{E} = \left(\frac{D_i - D_e}{\mu_i + \mu_e}\right)\frac{\vec{\nabla} n}{n}, \tag{1.11}$$

and the combined ambipolar flux and diffusion coefficient can be written as

$$\vec{\Gamma}_a = -\left(\frac{\mu_i D_e + \mu_e D_i}{\mu_i + \mu_e}\right)\vec{\nabla} n, \tag{1.12}$$

$$D_a = \frac{\mu_i D_e + \mu_e D_i}{\mu_i + \mu_e}, \tag{1.13}$$

respectively.

The motion of the charged particles in a plasma is governed by the Coulomb interaction forces between the particles, the external fields that are applied to the plasma and any internal fields created as a result of the motion of charged particles. As the charged particles move, any local concentration will give rise to internal electric fields. Moreover, their motion can lead to currents which generate internal magnetic fields. Both electric and magnetic fields can also be applied from external sources to the plasma. The motion of the charged particles in the plasma is therefore a result of all these external and internal forces and fields and the motion of each particle is as per Newton's laws of motion. Generally, the momentum of the plasma particles is high and therefore the de Broglie wavelength is small as compared to the interparticle distance (which goes as $n^{-1/3}$, where n is the plasma density). Hence, most plasmas can be treated classically except when the densities are very high or the temperatures are very low. The Lorentz force describes the interaction of the charged particles with the electromagnetic fields:

$$\boldsymbol{F} = q(\boldsymbol{E} + \mathbf{v} \times \boldsymbol{B}). \tag{1.14}$$

Therefore, the motion of the individual particles can be described by the above equation under the combined influence of the external and internal fields generated by all the plasma particles. The $\boldsymbol{E}$ and $\boldsymbol{B}$ fields obey Maxwell's equation. This method can sometimes become cumbersome to solve as there are a large number of particles. In case of any net charge in some plasma region, the potentials can be obtained by solving Poisson's equation:

$$\nabla^2\varphi = \frac{\rho}{\epsilon_o}. \tag{1.15}$$

Among the available theoretical treatments, depending upon the magnitude of the ratio of the particle mean free path to the system size (Knudsen number) two approaches can be considered to describe the transport of mass, momentum, and energy. At high values of the Knudsen number which also represents a case when the plasma densities are small one uses the single-particle approach also known as particle orbit theory which consists of studying the motion of each charged particle in the presence of specified electric and magnetic fields. Then there is the kinetic theory of a collection of such particles, which describe plasmas microscopically by means of particle distribution functions. And at low values of the Knudsen number or very high densities, the fluid approach may be more appropriate, where the plasma is described in terms of averaged macroscopic functions of position r and time t. The physics of the plasmas have been investigated through experimental, theoretical, and computational methods.

1.6 Plasma Sheaths and Generation of Ion Beams

When a material surface is introduced into a plasma, a sheath is formed in front of the surface on the plasma side. The sheath thickness is usually a few times the Debye length. Usually when a plasma is confined by walls, the electrons which are more mobile and have higher thermal velocities leave the bulk plasma very quickly and hit the wall, where they recombine and get lost. The bulk plasma as a result starts attaining net positive charge, and a potential which is positive with respect to the wall which is at a lower potential and negative. The potential between the wall and the plasma cannot be distributed in the entire plasma because Debye shielding will confine the potential variation within a layer of the order of a few Debye lengths. This layer will therefore separate the walls confining the plasma from the bulk plasma where quasi-neutrality is maintained and the layer is known as plasma sheath. The sheath therefore forms a potential barrier where the more mobile species, i.e., the electrons are confined electrostatically. The height of the potential barrier gets adjusted in a way that the flux of electrons that have enough energy to overcome the barrier is equal to the flux of ions reaching the wall. It is found that the ions enter the edge of the sheath from the plasma side with a velocity known as the Bohm velocity $V_B = \left(\frac{K_B T_e}{M}\right)^{1/2}$, where M is the mass of the ions. In a low-pressure plasma, the mean free path is usually much larger than the device dimensions, and therefore the ions that move through the sheath to hit the wall do not suffer from any collisions. Therefore, if an aperture is made in an electrode immersed in the plasma, the ions can be extracted out from the aperture as an ion beam. The ions would arrive at the aperture with Bohm velocity and then if an extraction voltage V is applied to an extractor electrode kept at a distance d away from the plasma electrode, the final velocity of the ion beam at the extractor electrode would be [8].

$$\mathrm{v_f} = \left(\mathrm{v_B^2} + \frac{2qV}{M}\right)^{1/2}, \tag{1.16}$$

where q is the charge of the ions. The ion beam current at the sheath edge would be decided by the maximum plasma density n_o and the maximum extractable current density at the aperture would be given by $J = en_o\mathrm{v_B}$. When a variable voltage V is applied at the extractor electrode, the extractable current density is given by the well-known Child-Langmuir's law [9].

$$J = \frac{4}{9}\left(\frac{2e}{M}\right)^{1/2}\frac{\varepsilon_o V^{3/2}}{d^2}, \tag{1.17}$$

where M is the mass of the ion, e is the charge of an electron, d is the distance between the plasma electrode and the extractor electrode, V is the applied voltage at the extractor electrode, and ε_o is the permittivity of free space. By looking at the above equation it would seem that as V is increased J can be increased indefinitely. However, the maximum value of J that can be attained is limited by the plasma density at the source and the Bohm velocity. The ion beams can therefore be extracted from the plasma source.

Questions

1. What are the constituents of a plasma? What is one distinguishing feature among the constituents?
2. Calculate the coupling parameter for a low-pressure hydrogen plasma where the electron temperature is 10 eV and the plasma density is 10^{18} m^{-3}.
3. Calculate the ionization fraction α for a hydrogen plasma at a temperature of 1 eV for number densities of 10^{15}, 10^{18}, 10^{21}, and 10^{24} m^{-3}.
4. Can any collection of charged particles be called as a plasma? What are the necessary conditions of a system of charged particles to be called as a plasma?
5. Calculate the screened potential at a distance of 5 λ_D for a plasma having a density of 10^{18} m^{-3} and a temperature of 10 eV.
6. Calculate the Debye length and plasma frequency for a plasma having a density of 10^{18} m^{-3} and a temperature of 5 eV.
7. Calculate the wavelength of the incident photons required to ionize Argon gas having an ionization potential of 15.76 eV.
8. Calculate the magnetic field required to realize the ECR condition for waves of 2.45 GHz.
9. What does ambipolar diffusion in plasma mean?
10. Calculate the Bohm velocity near a sheath edge for an Argon plasma having an electron temperature of 10 eV. How does this velocity compare with the average velocity of electrons in the plasma obtained by equating $\frac{1}{2}m\mathrm{v}^2 = \kappa_B T_e$.

References

1. H. Kählert, M. Bonitz, Dynamic structure factor of the magnetized one-component plasma: crossover from weak to strong coupling. Phys. Rev. Res. **4**, 013197 (2022)
2. J. A. Bittencourt, *Fundamentals of Plasma Physics*, Chap. 7 (Springer, 2004), p. 187
3. S. Hunjan, S. Bhattacharjee, Measurement of electron energetics in the equatorial and polar planes of a compact dipole plasma driven at steady state. Plasma Sources Sci. Technol. **32**, 105018 (2023)
4. S. Bhattacharjee, H. Amemiya, Microwave plasma in a multicusp circular waveguide with a dimension below cutoff. Jpn. J. Appl. Phys. **37**, 5742 (1998)
5. D. Garnier, A. Hansen, M. Mauel, E. Ortiz, A. Boxer, J. Ellsworth, I. Karim, J. Kesner, S. Mahar, A. Roach, Production and study of high-beta plasma confined by a superconducting dipole magnet. Phys. Plasmas **13**, 056111 (2006)
6. S. Bhattacharjee, A.R. Baitha, A. Nanda, S. Hunjan, S. Bhattacharjee, Physics of plasmas confined by a dipole magnet: insights from compact experiments driven at steady state. Rev. Mod. Plasma Phys. **6**, 16 (2022)
7. F.F. Chen, *Introduction to Plasma Physics and Controlled Fusion*, Chap. 5, vol. 1 (Springer, 2006)
8. S.K. Maurya, S. Paul, J.K. Shah, S. Chatterjee, S. Bhattacharjee, Momentum transfer using variable gaseous plasma ion beams and creation of high aspect ratio microstructures. J. Appl. Phys. **121**, 123302 (2017)
9. S. Bhattacharjee, A. Vartak, V. Mukherjee, Experimental study of space-charge-limited flows in a nanogap. Appl. Phys. Lett. **92**, 191503 (2008)

Chapter 2
Plasma Ion Sources and Beam Extraction from Plasmas

Various types of plasma-based ion sources have been developed worldwide [1–13] to suit diverse applications. These versatile sources can generate ions of a wide range of elements. Alongside plasma-based approaches, liquid metal (e.g., gallium) or liquid alloy sources have also been used for focused ion beam applications [14–18]. However, the latter can suffer from limitations, such as lower current values and the introduction of surface contamination, which could lead to short circuits in microelectronic devices. Notable plasma ion sources include Penning, filament-driven, RF-driven, and microwave-driven designs. Penning sources offer high current density but face issues with discharge instability and lower efficiency. Filament-driven sources have lower current density and shorter lifetimes due to filament wear and tear. RF-driven sources provide higher current density and smaller energy spread but can exhibit instability during the transition from capacitive to inductive coupling [9, 12]. To address these limitations, microwave-driven plasma ion sources have been developed, delivering stable high-density plasmas and high-current beams [13]. The advantages of plasma-based ion sources are manifold: (i) they can generate ion beam from gaseous plasmas of different elements, both noble/monatomic (He, Ar, Kr, Ne, Xe) and molecular (O_2, N_2, H_2); (ii) they can produce large beam currents ($\sim$1 pA–10 μA), and also in the mA range for example in high current ECR ion sources. Such ion sources are suitable for focused ion beam applications without causing metallic contamination, unlike liquid metal or alloy beams [13, 19–21]; and (iii) they can be readily used to treat semiconductors and biomaterials using inert gas ions that do not form chemical compounds on the surface.

S. Bhattacharjee, *Atomically Heterogeneous Systems*,
https://doi.org/10.1007/978-981-95-9733-8_2

2.1 Plasma-Based Ion Sources

Plasma-based ion sources have become increasingly popular and are now widely developed worldwide [1–15, 19–23]. These sources can generate ion beams capable of addressing a variety of research applications, including rapid processing of materials due to the availability of higher beam currents. In particular, the use of inert gas ions from plasma is becoming crucial for processing biomaterials, such as for artificial heart valves and orthopedic prostheses [24]. Additionally, ion milling applications employing a range of ion beam masses allow for controlled momentum transfer to the substrate, enabling the creation of microstructures with diverse aspect ratios [19, 20]. Various types of plasma ion sources are briefly described below.

2.1.1 Penning Ion Source

The formation of plasma in a Penning ion source is generally carried out using a cold cathode and high voltage. The oscillation of electrons between cathode and the anode leads to the generation of plasma which is confined by a Penning-type geometry [23]. To extract the ions from the plasma, an electrostatic lens, namely, Einzel lens are used. In spite of having a few complicacies, such as discharge instability, metallic contamination to the target sample, and low efficiency, these ion sources are advantageous due to compactness in design as well as capability of producing high current.

2.1.2 Filament-Driven Ion Source

The generation of plasma in a filament-based ion source is usually done by ionizing the gas via a filament and using a multicusp magnet to confine the plasma [11]. The primary advantage of such filament-type ion source is lesser spread in ion energy with controllability and universality. The major challenge in filament-type of ion source is a limited lifetime, therefore requiring frequent maintenance, contamination arising from the sputtered material, lower intensity of beam current, and restriction for generating plasma of corrosive gases.

2.1.3 RF Plasma-Driven Ion Source

Over the past 20 years, the radio frequency (RF) plasma-driven ion sources have been employed for diverse applications, such as material science including surface modification, space thruster, accelerator as well as fusion research. In addition, the RF plasma ion sources are used for several purposes such as plasma cleaning, ion beam etching, ion beam doping, micro-machining, and semiconductor fabrication [6]. The reduced cost of maintenance due to absence of filament is the primary advantage of such an ion source [7]. Besides, RF plasma ion sources have also been employed for the generation of negative ions for neutral beam injection in Tokamaks.

RF plasma ion sources are generated by the absorption of radio frequency (RF) energy, which creates the discharge. The primary advantage of RF-produced plasma is its ability to operate with a wide range of gases, including corrosive oxygen, which can significantly reduce the lifespan of a duoplasmatron used to generate a negative oxygen ion beam. Typically, the discharge chamber is maintained at a pressure between 10^{-3} and 10^{-2} mbar, and a few hundred watts of RF power are required to excite and sustain a stable plasma. For input power below several kilowatts, the widely used and commercially available frequency is 13.56 MHz. For high-intensity ion sources, such as those used in spallation neutron sources (SNS) or negative hydrogen ion sources, the RF frequency is generally 2 MHz, and the input power can reach up to 100 kW in pulsed mode [25]. There are two main types of RF ion sources: inductively coupled and capacitively coupled. Capacitively coupled plasma (CCP) is excited by a discharge between two electrodes applied with RF power, which generates an alternating potential across the electrodes. This CCP method is used to clean contaminated superconducting radio frequency (SRF) cavities [26] and ion source plasma chambers. The more common type of RF plasma and ion source is based on the inductively coupled method, also known as inductively coupled plasma (ICP). ICP is generated by an induction coil, where electrons in the gas are heated by an azimuthal electric field created by the alternating magnetic field around the induction coil, allowing them to acquire enough energy to form a plasma. To generate and sustain a dense, stable inductively coupled plasma, an RF antenna in the form of a coil is typically required. The RF antenna can be positioned either inside or outside the discharge chamber, commonly referred to as an internal or external RF antenna, respectively. For high-intensity, high duty cycle negative hydrogen ion sources, a multicusp magnet with an internal RF coil is widely employed [27, 28], and the copper-tube RF coil must be coated with a porcelain material. When the ion source operates at a high duty factor or in continuous wave (CW) mode, the RF antenna's lifetime may be affected, and high-voltage breakdown can occur with high-input RF power. As a result, the use of an external antenna is more extensively adopted for plasma and ion source studies. Typical antenna coils, which are predominantly cylindrical or planar in geometry, are illustrated in Fig. 2.1.

Kalvas et al. [29] developed a planar RF coil-driven ion source, and a milliampere H^- ion beam was extracted. The ion source design is a multicusp chamber (more details on the multicusp chamber will be discussed in the next section) with an external planar spiral RF antenna, behind a flat Aluminum Nitride (AlN) RF window on the back of the ion source. The maintenance interval reached 1 month, which was much longer than the filament-driven arc discharge ion source. Jiang et al. [30] developed a mini RF-driven ion source with 1.2 and 1.5 cm inner chamber diameters at Lawrence Berkeley National Laboratory, and operated it in the ICP mode. Different helical coupler antennas were tested, and high-brightness positive

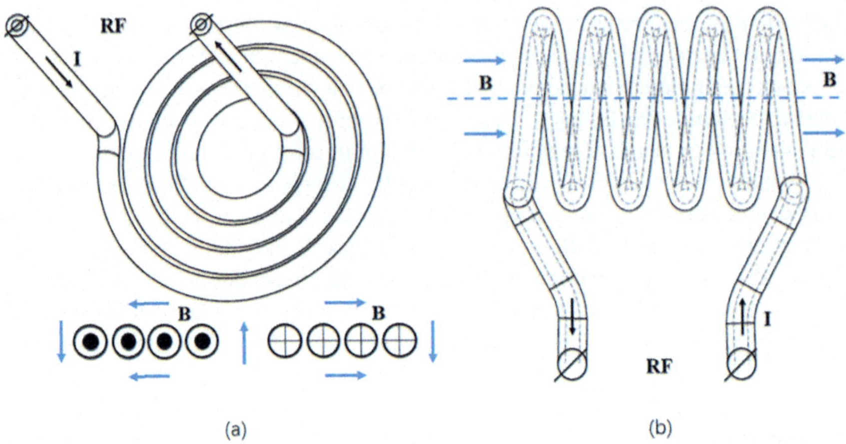

Fig. 2.1 Two RF antenna patterns used for ICP: **a** planar antenna, **b** cylindrical antenna (reprinted from [6] Plasma, under Creative Common CC BY license)

Ar^+, Kr^+, H^+ ions were extracted. The mini RF-driven ion sources have been utilized for producing nano-focused ion beams for focused ion beam (FIB) applications. Further Smith et al. [8] and Nabhiraj et al. [10] have separately developed RF ion sources for focused ion beam applications. These can provide good current density and small energy spread. Smith et al. have shown reduced brightness 5.4–9.1 × 10^3 Am^{-2} sr^{-1} V^{-1} and energy spread of a few eV for argon and xenon, respectively, and Nabhiraj et al. have produced argon ion beams with energy spread of less than 5 eV and brightness of 7100 Am^{-2} sr^{-1} eV^{-1}. However, as mentioned above, RF ion sources suffer from limitations such as power coupling issue, reduction of source lifetime, discharge transitions from capacitively coupled mode to inductively coupled mode, while changing input power or gas pressure, which therefore can interfere with the plasma properties and affect the beam extraction and milling performance.

2.1.4 Microwave Plasma Ion Sources

In microwave (MW) plasma ion sources, the plasma is generated using microwave energy. The seed electrons absorb this energy, leading to an increase in their kinetic energy through elastic collisions with neutral particles. Inelastic collisions then result in ionization, creating the plasma [31, 32]. Plasma confinement in these systems can be achieved in various ways, such as employing electromagnetic coils as in linear mirror machines, or utilizing a magnetic multicusp made of permanent magnets. The magnetic mirror effect in the cusp regions of the multicusp (near the boundary) leads to the formation of a dense central plasma [33]. Microwave plasmas offer several advantages over other ion sources. Since they do not require electrodes or filaments to sustain the plasma, issues related to sputtering and contamination are minimized. Additional benefits include: (i) a clean, contamination-free plasma; (ii) the ability to produce diverse ion beam species; (iii) high beam currents; (iv) an unmagnetized central beam extraction region using a multicusp plasma confinement device; (v) long lifetime and reduced maintenance requirements; and (vi) a narrow ion energy spread near the plasma meniscus, as observed in a microwave plasma-based focused ion beam system developed in our laboratory [34–40]. Microwave plasma ion sources often leverage the concept of electron cyclotron resonance (ECR) to generate high-density, high-electron-temperature plasma. In the presence of a magnetic field, this occurs when the wave frequency matches the cyclotron frequency of the electrons, enabling maximum coupling of wave energy to the particles [1–5]. The ECR magnetic field is typically generated using electromagnetic coils arranged in a magnetic mirror configuration, similar to a Helmholtz coil. Here the magnetic field is axial. The magnetic mirror ratio—the ratio of the maximum magnetic field (near the coils) to the minimum field (between the coils)—can be adjusted by varying the coil currents. This magnetic mirror ratio is an important parameter for ECR ion sources. In addition to the electromagnetic coils, many ECR ion sources worldwide also incorporate a magnetic multicusp field at the center to radially confine the plasma at the center. This multicusp is usually in a compact Halbach arrangement. Sometimes, for a more compact and simplified ECR-based microwave plasma ion sources, electromagnetic coils are omitted. Instead, the ECR magnetic field is realized near the boundary of a permanent magnet multicusp configuration, which provides the necessary plasma confinement [41–44]. Overall, the strategic use of ECR phenomena, magnetic mirror fields, and multicusp confinement enables the generation of high-performance microwave plasma ion sources used as injectors in cyclotrons, linear accelerators, and other applications (Figs. 2.2 and 2.3).

There exists several other types of ion sources specifically designed for extracting focused ion beams, one such example is a Cs^{+}-based low-temperature ion source (LoTIS). These type of ion sources are still under development and aimed for producing sub-nanometer beam spot with a magnitude of current ranging from 1 pA to few nano-amperes [46]. It has applications in the area of integrated secondary ion mass spectrometry [47]. A comparison of several types of ion sources is presented in Table 2.1. Apart from this there also exists other types of ion sources such as laser ion sources [48], electron impact gas ion source [49], electron beam ion sources [50], and ion traps [51, 52]. The laser ion sources are used to generate plasma by laser-induced ablation of target material, which is useful for both gases and metals.

2.2 Compact Multicusp Plasma Systems as an Ion Source

Plasmas produced through electron cyclotron resonance have been extensively studied in various research investigations [53–57]. The typical dimension of the plasma in systems utilizing coils or permanent magnets is around 10 cm. It is advantageous to generate plasma with a narrow cross section and high density for producing high-current, pencil-shaped ion beams, especially for ion source applications. These

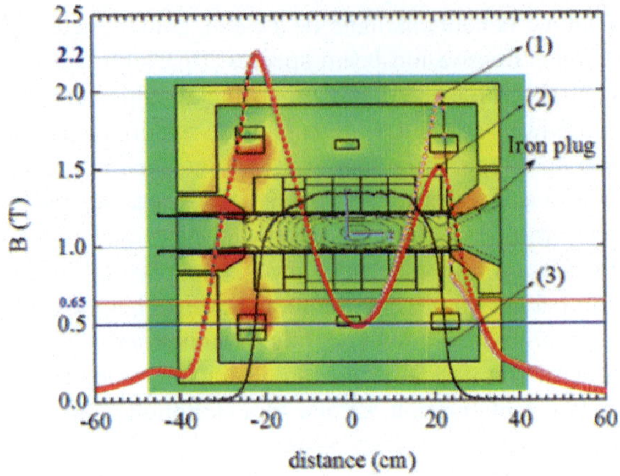

Fig. 2.2 The axial and the radial magnetic field in an ECR ion source created by four superconducting coils and the permanent magnet hexapole. (1) Axial magnetic field with the iron plug, (2) axial magnetic field without the iron plug, and (3) radial magnetic field by the hexapole (reprinted from [45] [H.-J. You, S.-W. Jang, Y.-H. Jung, T.-H. Lho, S.-J. Lee, *Rev. Sci. Instrum.* 83, 02A326 (2012)], with the permission of AIP Publishing)

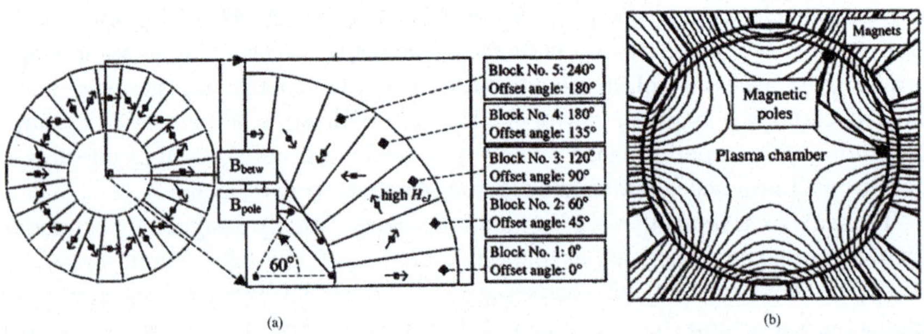

Fig. 2.3 Typical design of a Halbach-type hexapole and its magnetic field (reprinted from [45] [P. Suominen, O. Tarvainen, H. Koivisto, D. Hitz, *Rev. Sci. Instrum.* 75, 59–63 (2004)], with the permission of AIP Publishing)

Table 2.1 Source and beam parameters for different types of ion sources [8, 10, 18, 20, 23, 47] (reprinted from [20] [S. K. Maurya and S. Bhattacharjee, *Plasma Res. Express*, 2, 033001 (2020)], with the permission of IOP Publishing)

Ion sources	Reduced brightness (Am^{-2} sr^{-1} V^{-1})	Ion energy spread (eV)	Beam current
GFIS	1×10^9	1	0.1 pA to 10 pA
LMIS	1×10^6	5	0.2 pA to 20 nA
Penning	1×10^5	3	0.7 nA to 5 nA
ICP	4×10^2–1×10^4	3.5–5	0.1 pA to 10 μA
MWP	1×10^5	5	5 pA to 10 μA
LoTIS	2×10^7	0.34	0.5 pA to >1 nA

characteristics benefit ion sources in two key ways: (i) the higher plasma density enhances the ion beam current and (ii) the reduced cross-sectional size of the plasma chamber minimizes the beam's spot size, as the beam source size directly determines the object size. One approach for achieving higher plasma density involves decreasing the cross-sectional radius of the plasma chamber (typically circular), which increases the wave power density. This, in turn, enhances ionization within the chamber, resulting in denser plasma. However, the cross-sectional size of the waveguide is constrained by the geometrical wave cutoff condition, which governs the propagation of microwaves and the subsequent generation and maintenance of the plasma. Although there exists several literature, demonstrating methods to overcome both the geometrical [58, 59] and density cutoff limitations [59], these are predominantly focused on cases where the wave vector (K) is parallel to the static magnetic field B, i.e., K||B. In such scenarios, the geometrical cutoff condition is generally absent due to the propagation of the Whistler mode. However, experimental investigations [41–44, 60–62] have shown that in cases where K⊥B, such as when microwaves are introduced into a multicusp configuration, the waves can generate plasma near the entrance of the multicusp. This initial plasma alters the refractive index of the medium, enabling the waves to propagate with shorter wavelengths deeper into the multicusp, particularly through regions of upper hybrid resonance, as elaborated in the following section.

A detailed view of multicusp waveguide is depicted in Fig. 2.4 where the minimum radius is 3.4 cm and the length of the multicusp is 37 cm. In case of TE11 mode of circular waveguide, the magnitude of radius R should be lesser than the critical radius Rc (i.e., 3.6 cm). The fabrication of multicusp is made in such a way that it consists of 10 poles composed of rectangular pipes. The rectangular pipes consist of eight pieces of permanent magnets, each of dimensions $1.8 \times 1.5 \times 4.5$ cm^3. The poles of the magnets at both ends were oriented opposite to those of the central magnets to achieve magnetic plugging, at the end of each pipe. A cylindrical structure for the multicusp configuration provides stability in plasma generation achieved by securing the pipes to circular aluminum rings at both ends.

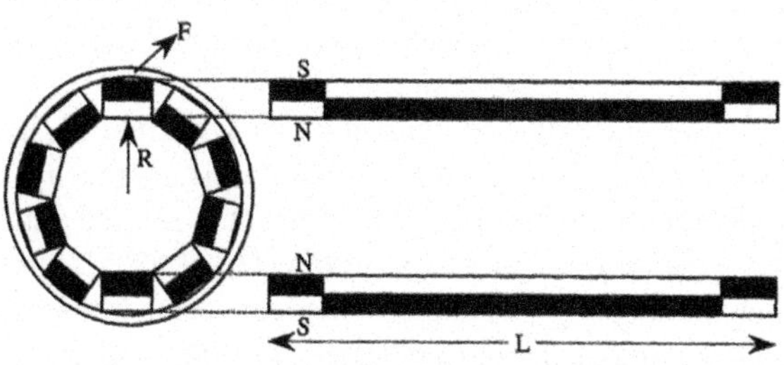

Fig. 2.4 Transverse and longitudinal sections of the multicusp waveguide MC. The inner radius R is 3.4 cm and the length L of the waveguide is 37 cm. Black and white parts indicate N and S poles of the magnets, respectively (reprinted from [20] [Bhattacharjee S and Amemiya H, Jpn J. Appl. Phys., 37, 5742 (1998)], with the permission of IOP Publishing)

2.2.1 Near-Boundary Resonances and Central Quiescent Plasma

The underlying mechanisms responsible for plasma generation within a narrow waveguide are explored in detail below. A representative example illustrating the properties of the plasma produced inside a multicusp device with dimensions smaller than the cutoff value is presented in the accompanying plot. The figure highlights how plasma can be sustained under conditions that traditionally limit wave propagation, shedding light on the unique dynamics at play within such a constrained geometry [41].

For the sake of convenience, three core regions, i.e., regions I ($r < 2.0$ cm), II ($2.0 < r \leq 2.65$ cm) and III ($2.65 < r < 3.4$ cm), are defined here. From a theoretical perspective, when the radial dimension of a waveguide is smaller than the cutoff value, electromagnetic waves are not expected to propagate through it. However, plasma ignition might still occur near the entrance of the waveguide due to the penetration of electric fields into this region. If microwaves manage to traverse the narrow section (Region II), the resulting microwave power density would significantly increase. This enhanced power density creates favorable conditions for plasma generation within the constricted area. At low neutral gas pressures, the plasma can diffuse into the waveguide, where it is effectively confined by the minimum-B configuration. This configuration minimizes radial losses, allowing the plasma to sustain itself within the restricted geometry. Additionally, the plasma potential (V_s) in Region I is observed to be higher than the ionization potential of the gas, as illustrated in Fig. 2.6. This high potential enables electrons to be accelerated from the peripheral boundary (Region III) toward the core (Region I), leading to enhanced ionization within the plasma. The elevated (V_s) also plays a crucial role in confining electrons in both radial and axial directions. This confinement reduces electron losses and increases the likelihood of collisions, further supporting ionization and maintaining the plasma within the waveguide. These combined effects illustrate how plasma can be generated and sustained in waveguides with dimensions below the cutoff value. As a result of these effects, the plasma density is greater than N_c, the cutoff density in the core region for microwave frequency 2.45 GHz. The magnitude of N_c is approximately 7.44×10^{10} cm^{-3} [63, 64].

The effective radial diffusion coefficient D_r is governed by Larmor radius in region III within the pressure range when a microscopic approach is considered. Let us assume that the longitudinal diffusion coefficient is D_z, the axial density $N+(z)$ can be estimated by $N+(z) = N +_o\exp[-(c/R)\ (D_r/\ D_z)^{0.5}\ z]$, where N_{+o} denotes the density at the entrance ($z = 0$) and c being a constant which can be calculated using the expression of radial distribution of N+ (i.e., Bessel *Jo*-type distribution with magnitude $c = 2.4$ for the lowest mode). If the classical or Bohm-type diffusion is considered, then the equation becomes $D_r/D_z \sim 1/(\omega\tau)^2$ or $0.06/(\omega\tau)$, where τ represents the mean collision time. For $p = 4 \times 10^{-4}$ Torr, $T_e = 6$ eV, and $B = 10^3$ G, the obtained $\omega\tau \sim 6 \times 10^3$. $N+(z)/N +_o$ is ~0.99 and ~ 0.92 for the entrance and exit ($z = 37$ cm), respectively. The measured density profiles, as depicted in Fig. 2.7, reveal a subtle increase in plasma density along the (z)-axis as one approaches the central region of the multicusp device. This observed trend suggests that, in addition to plasma diffusion through the waveguide, there is a supplementary generation of charged particles within the system. This enhancement can be attributed to the mechanisms associated with the additional ionization due to the plasma potential. The high (V_s) facilitates the acceleration of electrons toward the plasma's core region, thereby increasing ionization rates and contributing to the overall plasma density. This process, combined with the intrinsic properties of the multicusp magnetic field, highlights the device's ability to support not only plasma confinement but also the effective production and sustenance of charged particles. The multicusp magnetic field plays a critical role in this context, acting as a confinement structure that minimizes radial and axial particle losses while simultaneously promoting the diffusion and retention of the plasma. Thus, the interplay between the magnetic field configuration and the high (V_s) conditions synergistically enhances the density and stability of the plasma, as evidenced by the density profile trends observed in the experimental measurements.

The comparison of plasma in tubes of square and circular cross sections [42] is presented in Fig. 2.5. The rate of ionization, considering the electron temperature T_e, and N_{ew} by considering the space charge potential energy were calculated using the expressions $N_e\sigma(\kappa_B T_e)v_e$ and $N_{ew}\sigma(eV_s)\left(\frac{2eV_s}{m_e}\right)^{\frac{1}{2}}$, respectively, as shown in Fig. 2.8. Here N_e is the electron density, N_{ew} is the electron density near the wall ($N_{ew}/N_e \sim 1/10$), σ is the ionization cross section, v_e is the electron thermal velocity, and m_e is the electron mass. An assumption that the electrons near the wall were accelerated to the core plasma with an energy eV_s and cause ionization was considered to explain the potential ionization effect. The ionization effects of V_s and T_e can be compared at lower magnitude of power P_w, whereas scenario changes with dominating effect of T_e over V_s are observed at higher magnitude of power. Such type of tendency is found in circular tube. The production as well as maintenance of the overdense plasma in the case of the circular tube is thus aided by the effect of ionization for both V_s and T_e although the latter effect is higher as the microwave power is increased (Fig. 2.9).

The plasma created within a tube is a perfect example of an inhomogeneous and anisotropic medium which disrupts the typically existing well-defined wave modes. However, the upper hybrid resonance (UHR) plays a key role in production and maintenance of plasma governed by the equation $\alpha^2 + \beta^2 = 1$, where $\alpha = \omega_p/\omega$ and $\beta = \omega_p/\omega$. Here, ω_p, ω_c, and ω represent the angular plasma, electron cyclotron, and wave frequencies, respectively. The change in refractive index within the magnetized regions allows waves to propagate through the periphery of the central over dense plasma, near the UHR regions, with a relatively short wavelength compared to the tube dimensions. The wavelength of the waves could be estimated from:

$$\left(\frac{k}{k_o}\right)^2 = \left(\frac{\lambda_o}{\lambda}\right)^2 = 1 - \frac{\alpha^2[(1-j\delta)-\alpha^2]}{(1-j\delta)^2-\beta^2-\alpha^2(1-j\delta)} - \left(\frac{k_c}{k_o}\right)^2, \tag{2.1}$$

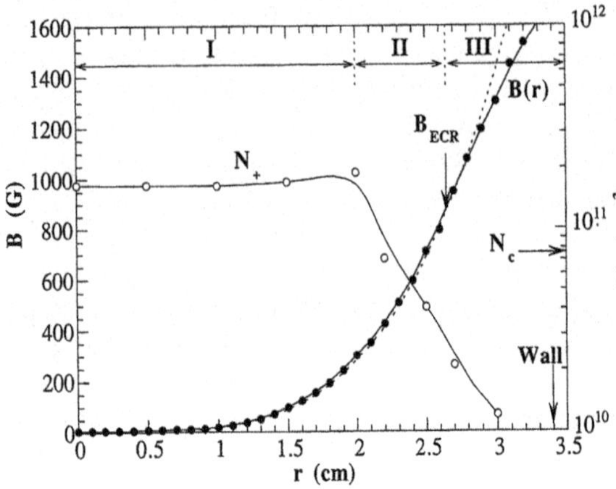

Fig. 2.5 Radial variations of the magnetic field B and plasma ion density N+ at power $P_f = 300$ W and pressure $p = 2.8 \times 10^{-4}$ Torr (reprinted from [20] [Bhattacharjee S and Amemiya H, Jpn J. Appl. Phys., 37, 5742 (1998)], with the permission of IOP Publishing)

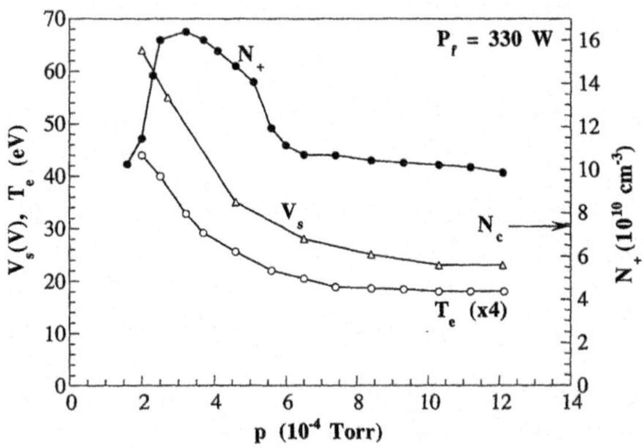

Fig. 2.6 Variations of the plasma (ion) density N+, space potential V_s, and electron temperature T_e with pressure p for $P_f = 330$ W (reprinted from [20] [Bhattacharjee S and Amemiya H, Jpn J. Appl. Phys., 37, 5742 (1998)], with the permission of IOP Publishing)

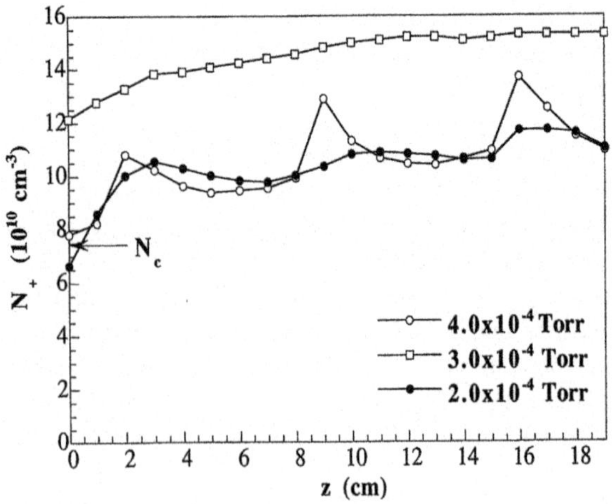

Fig. 2.7 Axial variation of the plasma ion density N+ from one end to the middle of the MC. Measurements for pressures $p = 2.0 \times 10^{-4}$ and 3.0×10^{-4} Torr were made at power $P_f = 300$ W and for $p = 4.0 \times 10^{-4}$ Torr at $P_f = 260$ W (reprinted from [20] [Bhattacharjee S and Amemiya H, Jpn J. Appl. Phys., 37, 5742 (1998)], with the permission of IOP Publishing)

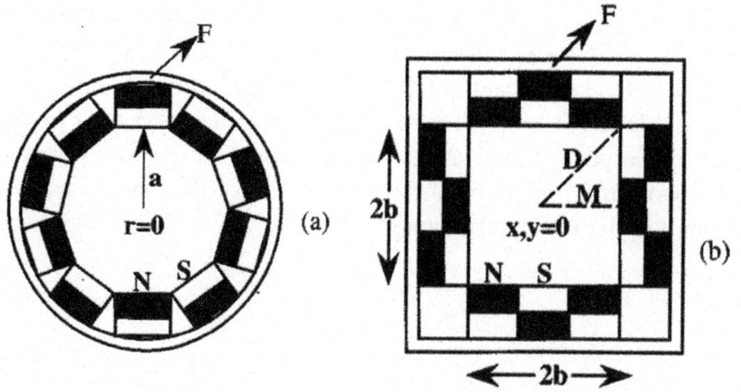

Fig. 2.8 Schematic of the cross sections of the: **a** circular and **b** square tubes. The radius of the circular tube $a = 3.4$ cm and the side $2b$ of the square tube = 6 cm. M and D are directions along the midline and the diagonal of the square tube, respectively. F is an aluminum frame. The tubes are so designed that the cross-sectional areas are approximately the same and the length L being same, this ensures that the microwave power density is approximately the same (reprinted from [42] [Bhattacharjee S and Amemiya H, Rev. Sci. Instrum., 70, 3332 (1999)], with the permission of AIP Publishing)

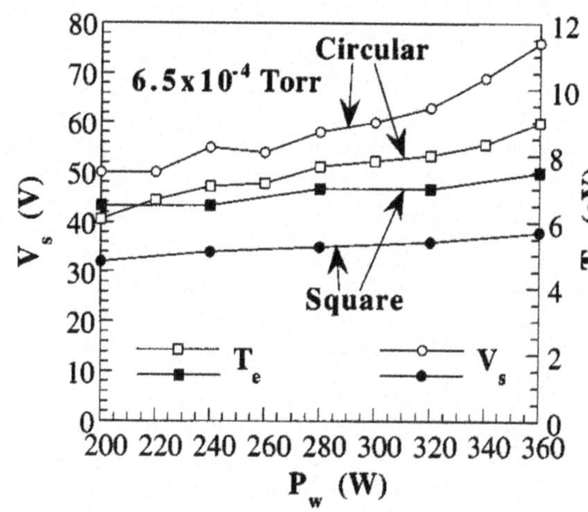

Fig. 2.9 Variations of the space potential V_s and electron temperature T_e with the feed power Pw at a pressure of 6.5×10^{-4} Torr for the circular and square tubes, respectively (reprinted from [42] [Bhattacharjee S and Amemiya H, Rev. Sci. Instrum., 70, 3332 (1999)], with the permission of AIP Publishing)

where k, k_o, and k_c are the wave numbers in the magnetized plasma in a waveguide, that in free space, and the cutoff wave number in the waveguide without plasma, respectively. $\delta = \nu/\omega$, where ν is the electron neutral collision frequency. As a typical example, the wavelength near the UHR region has been calculated using the experimental parameters, $k_o = 0.5133$ cm^{-1}, $k_c = 0.5417$ cm^{-1} (circular tube), $\lambda_o =$ 12.24 cm, and $\delta \leq 10^{-3}$. For $B = 500$ G, $N = 5.0 \times 10^{10}$ cm^{-3}, we obtain $\lambda \sim$ 0.02 cm. Propagation of high-frequency waves through the peripheral region of the tube can serve to maintain the plasma discharge.

Thus, microwave plasmas can be produced within a narrow waveguide, enabling the creation of over dense plasmas with densities exceeding the cutoff value. This discovery holds significant benefits for the development of ion sources from microwave-produced plasmas, as it facilitates the generation of a pencil-like ion beam from the compact plasma source. Appropriate electrostatic Einzel lenses are then employed to extract the ion beam from the plasma. These aspects will be explored in greater detail in the following chapters.

Questions

1. List the different types of ion sources that is based upon operation using DC voltages and electromagnetic waves.
2. What are the main issues in DC filament-based plasma generation and ion sources?
3. What are the principal advantages of microwave plasmas and MW plasma-based ion sources?
4. What are the two main methods by which RF plasmas are generated and what is the most widely used RF frequency? What are the main instabilities in RF plasmas?
5. Do an Internet search and list up the main accelerator laboratories in the World that employ electron cyclotron resonance ion sources (ECRIS), alongside write the names of the machines as well. What are the typical magnitudes of extracted ion beam currents and charge states in these ECRIS?
6. Can a beam of metal ions be generated from a plasma? Read about the generation of metal ions from organic metal compounds (MIVOC method) used in ECRIS.
7. What are the limitations of microwave propagation in narrow cross section waveguides? What does it depend upon?
8. Are there any limitations posed by the plasma density in the propagation of electromagnetic waves in a plasma?
9. How have the authors overcome both the waveguide geometrical cutoff and the plasma density cutoff limitations, in generating a high-density plasma in a narrow cross-sectional waveguide?
10. Why does a circular multicusp support a higher density than a square multicusp?
11. What are the key factors that contribute to the stability of the central quiescent plasma? Why does the motion of ions and electrons near the plasma boundary differ from their motion in the central region?

References

1. R. Geller, ECRIS: The electron cyclotron resonance ion sources. Annu. Rev. Nucl. Part. Sci. **40**, 15–43 (1990)
2. Y. Higurashi, T. Nakagawa, M. Kidera, T. Aihara, M. Kase, Y. Yano, Effect of plasma electrode position of RIKEN 18 GHz electron cyclotron resonance ion source on beam intensity of highly charged Ar ions. Jpn. J. Appl. Phys. **44**(7A), 5216–5218 (2005)
3. M. Kidera, M. Lamoureux, V. Mironova, G. Shirkov, Correlation between the hot-electron dynamics and the afterglow currents in electron cyclotron resonance ion sources. Rev. Sci. Instrum. **70**, 4234–4237 (1999)
4. G.S. Taki, D.K. Chakraborty, R.K. Bhandari, 6.4 GHz ECR ion source at VECC. PRAMANA—J. Phys. **59**(5), 775–780 (2002)
5. P. Kumar, G. Rodrigues, U.K. Rao, C.P. Safvan, D. Kanjilal, A. Roy, ECR ion source based low energy ion beam facility. PRAMANA—J. Phys. **59**(5), 805–809 (2002)
6. Q.Y. Jin, Y.G. Liu, Y. Zhou, W. Q., Y.J. Zhai, L.T. Sun, RF and microwave ion sources study at institute of modern physics. Plasma **4**, 332–344 (2021)
7. E. Speth, M. Ciric, J.H. Feist, P. Frank, B. Heinemann, W. Kraus, F. Probst, R. Riedl, R. Trainham, O. Vollmer, R. Wilhelm, Rf ion sources for fusion applications: Design, development and performance. Fusion Eng. Des. **46**, 383–388 (1999)
8. N.S. Smith, W.P. Skoczylas, S.M. Kellogg, D.E. Kinion, P.P. Tesch, O. Sutherland, A. Aanesland, R.W. Boswell, High brightness inductively coupled plasma source for high current focused ion beam applications. J. Vac. Sci. Technol. B **24**, 2902 (2006)
9. Q. Ji, X. Jiang, T.J. King, K.N. Leung, K. Standiford, S.B. Wilde, Improvement in brightness of multicusp plasma ion source. J. Vac. Sci. Technol. B **20**, 2717 (2002)
10. P.Y. Nabhiraj, R. Menon, G.M. Rao, S. Mohan, R.K. Bhandari, Characterization of compact ICP ion source for focused ion beam applications. Nucl. Instrum. Methods Phys. Res., Sect. A **621**, 57 (2010)
11. L. Scipioni, D. Stewart, D. Ferranti, A. Saxonis, Performance of multicusp plasma ion source for focused ion beam applications. J. Vac. Sci. Technol. B **18**, 3194 (2000)
12. S.K. Guharay, E. Sokolovsky, J. Orloff, Characteristics of ion beams from a penning source for focused ion beam applications. J. Vac. Sci. Technol. B **17**, 2779 (1999)
13. J.V. Mathew, I. Dey, S. Bhattacharjee, Microwave guiding and intense plasma generation at subcutoff dimensions for focused ion beams. Appl. Phys. Lett. **91**, 041503 (2007)
14. L.D. Menard, J.M. Ramsey, Fabrication of sub−5 nm nanochannels in insulating substrates using focused ion beam milling. Nano Lett. **11**, 512 (2011)
15. L.A. Giannuzzi, F.A. Stevie, A review of focused ion beam milling techniques for TEM specimen preparation. Micron **30**, 197 (1999)
16. Y. Hirayama, S. Tarucha, Y. Suzuki, H. Okamoto, Fabrication of a GaAs quantum-well-wire structure by Ga focused ion beam implantation and its optical properties. Phys. Rev. B **37**, 277 (1988)
17. J. Gupta, J.M.E. Harper, J.L. Mauer, P.G. Blauner, D.A. Smith, Focused ion-beam imaging of grain-growth in copper thin-films. Appl. Phys. Lett. **61**, 663 (1992)
18. J. Orloff, M. Utlaut, L. Swanson, *High Resolution Focused Ion Beams: FIB and its Applications* (Kluwer Academic, Plenum Publisher, New York, 2003)
19. S.K. Maurya, S. Paul, J.K. Shah, S. Chatterjee, S. Bhattacharjee, Momentum transfer using variable gaseous plasma ion beams and creation of high aspect ratio microstructures. J. Appl. Phys. **121**, 123302 (2017)
20. S.K. Maurya, S. Bhattacharjee, Variable gaseous ion beams from plasmas driven by electromagnetic waves for nano-micro structuring: a tutorial and an overview of recent works and future prospects. Plasma Res. Express **2**, 033001 (2020)
21. S.K. Maurya, S. Barman, N. Pan, S. Bhattacharjee, Effect of plasma and beam parameters on focal dimensions in micrometer charged particle optics: Enhanced nonlinear demagnification below the Debye length. Phys. Plasmas **26**, 063103 (2019)

22. S. Paul, A. Chowdhury, S. Bhattacharjee, Effect of beam limiting aperture and collector potential on multi element focused ion beams. Rev. Sci. Instrum. **83**, 02B714 (2012)
23. H.W. Loeb, Plasma based ion beam sources. Plasma Phys. Controlled Fusion **47**, B565 (2005)
24. F.Z. Cui, Z.S. Luo, Biomaterials modification by ion-beam processing. Surf. Coating Tech. **112**, 278–285 (1999)
25. I.G. Brown, *The Physics and Technology of ion Sources*, 2nd edn. (Wiley, Weinheim, 2004), pp. 163–175
26. A. Wu, L. Yang, C. Hu, C. Li, S. Huang, Y. Li, Q. Chu, P. Xiong, H. Guo, W. Yue, et al., In-situ plasma cleaning to decrease the field emission effect of half-wave superconducting radio frequency cavities. Nucl. Instrum. Methods Phys. Res. A **905**, 61–67 (2018)
27. M.P. Stockli, B. Han, M. Clemmer, S.M. Cousineau, A. Justice, Y.W. Kang, S.N. Murray, T.R. Pennisi, C. Piller, C.M. Stinson, et al., Upgrading the LANSCE accelerator with a SNS RF-driven H− ion source. Rev. Sci. Instrum. **91**, 013321 (2020)
28. K. Shinto, K. Ohkoshi, T. Shibata, K. Nanmo, K. Ikegami, A. Takagi, Y. Namekawa, A. Ueno, H. Oguri, Progress of the J-PARC cesiated RF-driven negative hydrogen ion source. AIP Conf. Proc. **2052**, 050002 (2015)
29. T. Kalvas, O. Tarvainen, J. Komppula, H. Koivisto, J. Tuunanen, D. Potkins, T. Stewart, M.P. Dehnel, A CW radiofrequency ion source for production of negative hydrogen ion beams for cyclotrons. AIP Conf. Proc. **1655**, 030015 (2015)
30. X. Jiang, Q. Ji, A. Chang, K.N. Leung, Mini rf-driven ion sources for focused ion beam systems. Rev. Sci. Instrum. **74**, 2288–2292 (2003)
31. S. Bhattacharjee, S. Paul, Random walk of electrons in a gas in the presence of polarized electromagnetic waves: Genesis of a wave induced discharge. Phys. Plasmas **16**, 104502 (2009)
32. S. Bhattacharjee, I. Dey, S. Paul, Electron random walk and collisionsl crossover in a gas in the presence of electromagnetic waves and magnetic fields. Phys. Plasmas **20**, 042118 (2013)
33. J.A. Bittencourt, *Fundametals of Plasma Physics*, 3rd edn. (Springer, New York, 2004)
34. J.V. Mathew, S. Paul, S. Bhattacharjee, Ion energy distribution near a plasma meniscus with beam extraction for multielement focused ion beams. J. Appl. Phys. **107**, 093306 (2010)
35. J.V. Mathew, Multi-element focused ion beams from intense microwave plasmas. Physics and applications thesis. Department of Phys. IIT Kanpur (http://172.28.64.70:8080/jspui/handle/123456789/12039) (2010)
36. J.V. Mathew, S. Bhattacharjee, Compact electrostatic beam optics for multi element focused ion beams: Simulation and experiments. rev. Sci. Instrum. **82**, 013501 (2011)
37. S. Bhattacharjee, S. Paul, Genesis of focused ion beams for plasma nanotechnology using a bounded microwave plasma source. Jpn. J. Appl. Phys. **54**, 01AA06 (2015)
38. J.V. Mathew, A. Chowdhury, S. Bhattacharjee, Subcutoff microwave driven plasma ion sources for multielemental focused ion beam systems. Rev. Sci. Instrum. **79**, 063504 (2008)
39. J.V. Mathew, S. Bhattacharjee, Multi-element focused ion beams using compact microwave plasma ion source. Indian J. Phys. **85**, 1863 (2011)
40. S. Paul, Micron Focusing, Diagnostics and Structuring Using Multi-Element Ion Beams. Doctoral Thesis Department of Physics, Indian Institute of Technology-Kanpur, Kanpur (2015) (http://172.28.64.70:8080/jspui/handle/123456789/14772)
41. S. Bhattacharjee, H. Amemiya, Microwave plasma in a multicusp circular waveguide with a dimension below cutoff. Jpn. J. Appl. Phys. **37**, 5742 (1998)
42. S. Bhattacharjee, H. Amemiya, Production of microwave plasma in narrow cros-sectional tubes; effect of the shape of cross section. Rev. Sci. Instrum. **70**, 3332 (1999)
43. S. Bhattacharjee, *Compact Plasma and Focused Ion Beams* (CRC Press, Boca Raton, FL, 2014)
44. S. Bhattacharjee, H. Amemiya, Transversely magnetized microwave plasma in a rectangular waveguide under cutoff conditions. Rev. Sci. Instrum. **68**, 3061–3067 (1997)
45. P. Suominen, O. Tarvainen, H. Koivisto, D. Hitz, Optimization of the Halbach-type magnetic multipole for an electron cyclotron resonance ion source. Rev. Sci. Instrum. **75**, 59–63 (2004)
46. N.S. Smith, J.A. Notte, A.V. Steele, Advances in source technology for focused ion beam instruments. MRS Bull. **39**, 329–335 (2014). www.mrs.org/bulletin

47. S.K. Guharay, S. Douglass, J. Orloff, High-resolution primary ion beam probe for SIMS. Appl. Surf. Sci. **926**, 231–232 (2004)
48. B.G. Freinkman, A.V. Eletskii, S.I. Zaitsev, A proposed laser source of ions for nanotechnology. microelectron. Eng. **73–74**, 139–143 (2004)
49. J.E. Barth, C.B. de Gruyter, E. Koets, P. Kruit, P.E. van der Leeden, J.B. Le Poole, K.D. van der Mast, Gas ion source for focused beams microelectron. Engineering **3**, 147 (1985)
50. G. Zschornack, M. Schmidt and A. Thorn, Electron Beam Ion Sources, arxiv.org/abs/1410.8014
51. R.E. Marrs, Self-cooling of highly charged ions during extraction from electron beam ion sources and traps. Nucl. Instrum. Methods Phys. Res. B **149**, 182 (1999)
52. J.W. McDonald, R.W. Bauer, D.H.G. Schneider, Extraction of highly charged ions (up to 90+) from a high-energy electron beam ion trap. Rev. Sci. Instrum. **73**, 30 (2002)
53. T.D. Mantei, S. Dhole, Characterization of a permanent magnet electron cyclotron resonance plasma source. J. Vac. Sci. Technol. B **9**, 26 (1991)
54. S. Iizuka, N. Sato, Electron cyclotron resonance devices with permanent magnets for production of large-diameter uniform plasmas. Jpn. J. Appl. Phys. **33**, 4221 (1994)
55. M. Fukao, K. Matsuura, Overdense plasmas produced by electron cyclotron heating in a linear multipole. Jpn. J. Appl. Phys. **33**, 4232 (1994)
56. M. Sugimoto, M. Tanaka, Y. Kawai, Electron cyclotron wave plasma production using a concave lens. Jpn. J. Appl. Phys. **35**, 2803 (1996)
57. S.R. Douglass, C. Eddy Jr., B.V. Weber, Faraday rotation of microwave fields in an electron cyclotron resonance plasma. IEEE Trans. Plasma Science **24**, 16 (1996)
58. O.A. Popov, Characteristics of electron cyclotron resonance plasma sources. J. Vac. Sci. Technol. A **7**, 894 (1989)
59. M. Shimada, Y. Torii, Compact electron cyclotron resonance ion source with a permanent magnet. J. Vac. Sci. Technol. A **11**, 1313 (1993)
60. S. Bhattacharjee, H. Amemiya, Pulsed microwave plasma production in a conducting tube with a radius below cutoff. Vacuum **58**, 222 (2000)
61. S. Bhattacharjee, H. Amemiya, Production of microwave plasma in a waveguide with a dimension below cutoff. J. Tech. Phys. **40**, 213 (1999)
62. H.-J. You, S.-W. Jang, Y.-H. Jung, T.-H. Lho, S.-J. Lee, Design and fabrication of a superconducting magnet for an 18 GHz electron cyclotron resonance ion/photon source NFRI-ECRIPS. Rev. Sci. Instrum. **83**, 02A326 (2012)
63. M. Tanaka, R. Nishimoto, S. Higashi, N. Harada, T. Ohi, A. Komori, Y. Kawai, Overdense plasma production using electron cyclotron waves. J. Phys. Soc. Jpn. **60**, 1600 (1991)
64. S. Bhattacharjee, H. Amemiya, Production of pulsed microwave plasma in a tube with a radius below the cutoff value. J. Phys. D Appl. Phys. **33**, 1104 (2000)

Chapter 3
Generation of Broad Ion Beams and Focused Micro-ion Beamlets

3.1 Introduction

Over the decades, ion beam technology has evolved significantly, with ion energies ranging from a few electron volts (eV) to several mega-electron volts (MeV), enabling a diverse array of research and technological applications. The penetration depth of ions within a material is inherently dependent on their energy, allowing precise modification of specific layers by carefully selecting the beam energy. This tunability makes ion beams an indispensable tool in materials science, semiconductor fabrication, and nanotechnology. Furthermore, different ion species can be employed based on the desired application, ranging from inert noble gas ions for defect engineering to reactive ions for chemical modifications. Most ion beam applications utilize either broad beams or highly focused micro-beams to achieve the required modifications. In certain cases, the use of multiple micro-beamlets can offer enhanced capabilities by enabling localized, site-specific alterations. Historically, research and industrial applications have predominantly relied on single ion beams that are extracted and directed onto substrates. However, emerging technologies now demand multiple ion beamlets for advanced material structuring. This is particularly relevant in areas such as nanoscale surface patterning, functionalization of soft matter (e.g., polymers) [1], creation of engineered two-dimensional (2D) and three-dimensional (3D) surface architectures for nanoelectronics, spintronic and photonic devices, and bioengineering applications. Additionally, multi-ion beamlets play a crucial role in complex mask fabrication for nanoelectronics, bio-patterning, and mask-less lithography [2, 3]. The advantages of utilizing multiple ion beamlets are substantial. First, they allow for the direct fabrication of intricate patterns on substrates with high spatial resolution. Second, they enable localized doping and surface modification, such as controlled alterations in electrical resistivity at designated regions [4, 5]. A functional surface, in this context, refers to a substrate that has undergone targeted modifications to acquire new properties absent in its pristine form, such as enhanced electrical conductivity, modified optical absorption [6],

S. Bhattacharjee, *Atomically Heterogeneous Systems*,
https://doi.org/10.1007/978-981-95-9733-8_3

or tunable reflectivity [7]. Such precisely engineered surfaces are expected to meet the stringent demands of next-generation micro- and nanotechnology applications, which conventional, uniformly treated surfaces fail to fulfill. Despite the evident advantages, the controlled generation and application of multi-ion beamlets remain a largely unexplored area, with only a few reports addressing this challenge. A particularly noteworthy recent study demonstrated a novel approach using filament-generated plasmas [3] to achieve structured multi-ion beams. The development of a versatile system capable of delivering multiple focused ion beams of various elements would significantly expand the scope of ion beam applications, opening new frontiers in materials engineering, semiconductor processing, and beyond.

Recent advancements in low-energy ion beam physics have led to a diverse range of studies that can be broadly classified into three major research domains:

1. Ion-Induced Nano-Structure Formation

Under specific conditions, such as optimized beam incident angles and appropriate ion-target material pairings, ion beam sputtering can induce well-ordered ripple patterns on surfaces. These self-organized nanostructures have been observed on various substrates, including silicon (Si) [8, 9], glass [10], metals such as platinum (Pt) [11], amorphous carbon [12], and pyrochlore ($Cd_2Nb_2O_7$) [13]. Heavier noble gas ions (Ar^+, Kr^+, Xe^+) have been commonly employed to generate these nanoscale periodic patterns. In some cases, continuous ion irradiation leads to ripple propagation, as evidenced by scanning electron microscopy (SEM) studies. Notably, a 30 kV Ga^+ ion beam was used to induce and track the evolution of ripples on pyrochlore surfaces [13], with propagation velocities being quantitatively measured. This phenomenon has garnered significant research interest due to its potential applications across various domains: magnetic property manipulation for advanced spintronic devices [14]; controlled adsorption of large molecules, impacting molecular self-assembly and catalytic reactions [15]; and tuning the chemical reactivity of catalytic surfaces, enabling the engineering of selective chemical processes [16], including optoelectronic applications, where ripple-induced changes in surface morphology alter light-matter interactions [17].

2. Nano-Gas Bubble Formation in Metals

The formation of nanoscale gas bubbles within metals is another intriguing consequence of low-energy ion implantation. When rare-gas atoms are embedded into metallic matrices, they aggregate to form nanoscale gas-filled cavities, creating a fascinating nano-system [18, 19]. Reported bubble sizes range from sub-nanometer scales to less than 10 nm in diameter [20]. The unique physical properties of these rare-gas bubbles make them an exciting subject of investigation such as existence in the solid state at room temperature due to over-pressurization, extreme mass densities, and the potential to exist in a metastable superheated state [21]; quantum confinement effects, resulting from their nanometric size and close proximity to the metallic surface [22]; electron interference phenomena, where subsurface gas bubbles influence conduction electrons, leading to novel electronic behaviors [22]; and photoemission spectroscopy applications, where rare-gas atoms in nearly-free-electron metals serve as ideal systems for studying conduction electron responses in excited states [18].

3. Physico-Chemical and Electronic Property Modification

Low-energy ion beams offer a powerful tool for tailoring surface properties, allowing precise modifications in morphology, crystallographic structure, and electronic characteristics of materials. Controlled ion irradiation has been shown to induce significant changes in electrical [23–25], optical, and thermal properties. However, most prior studies have relied on single broad ion beams (diameter ≥ 0.1–2 cm), limiting the ability to achieve spatially localized modifications. This constraint has hindered progress in micron- and nanoscale surface engineering, where precise control over implantation is critical.

4. Advancements in Multi-ion Beamlet Technology

This chapter introduces a novel experimental approach utilizing micron-scale multiple ion beamlets to investigate the interaction of low-energy (1–5 keV) ion beams with various substrate materials at micrometer resolutions. The advantages of using multiple low-energy beamlets include high-resolution surface patterning, enabling the direct fabrication of nanoscale structures and localized property modification, allowing for the creation of site-specific resistive, conductive, or optically active regions. Enhanced control over implantation depth and lateral distribution, crucial for micron- and nanoscale doping applications in semiconductor and optoelectronic device fabrication. By developing and optimizing multi-ion beamlet systems, the present research paves the way for next-generation micron-to-nanoscale implantation technologies, where controlled surface structuring and functionalization play pivotal roles in emerging electronic and photonic applications.

3.2 Experimental Setup

An experimental setup to generate broad and focused micro-ion beamlets from a plasma source is shown in Fig. 3.1. Depending upon the type of plasma electrode (PE) employed in the setup, one can realize either broad or focused micro-ion beamlets. The detail of the setup is provided.

The schematic of the experimental setup is shown in Fig. 3.1. Plasma is produced inside a magnetic multicusp (MC) kept inside a vacuum chamber (VC). The MC is developed with permanent magnets which are arranged in a cylindrical geometry having alternate north and south poles such that a minimum-B field is created at the center. Figure 3.1b shows a 2D Poisson simulation of the magnetic field lines. Microwaves of 2.45 GHz are produced by a magnetron oscillator and are guided through an isolator, a directional coupler, and a triple stub tuner and launched using a straight waveguide (WG) into VC through a quartz window (W). VC is evacuated by a turbo molecular pump (TMP) to a base pressure of 10^{-7} Torr, backed by a diaphragm pump. The feed gas is introduced into the chamber through a gas inlet

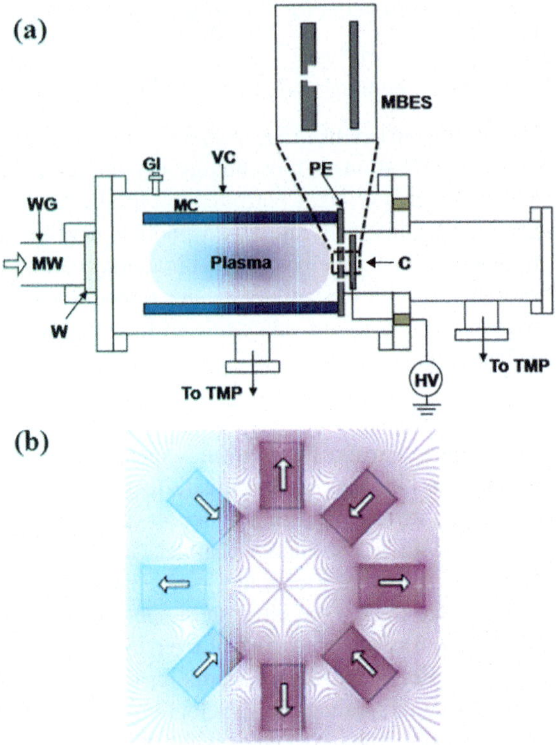

Fig. 3.1 **a** The experimental setup with all the components. *MW* microwaves, *WG* rectangular waveguide, *W* quartz window, *VC* vacuum chamber, *TMP* turbo molecular pump, *GI* gas inlet, *MC* multicusp, *PE* plasma electrode, *C* collector, *HV* high-voltage power supply, *MBES* multi-beam electrode system (inset: electrode geometry). **b** 2D Poisson simulations of multicusp (cross-sectional view) showing the magnetic field lines) (reprinted from [Abhishek Chowdhury and Sudeep Bhattacharjee, AIP Advances 1, 042150 (2011)], with the permission of AIP Publishing)

(GI). The gas pressure is controlled by a mass flow controller (MKS 1179A) and is maintained in the range of 0.085–0.55 mTorr. The multiple beamlet electrode system (MBES), which can help in the generation of multiple ion beamlets, is attached to a flange which is then attached to MC. For generation of broad ion beams, a single through-aperture electrode can also be attached as shown in Fig. 3.1. The plasma production in this system and its utilization as an ion source has already been demonstrated and discussed in the earlier chapters. To extract multiple ion beamlets, the MBES system which consists of a plasma electrode (PE) with a honeycomb-like pattern has been designed, it contains apertures at the vertices and at the center of a hexagon (as shown in Fig. 3.2b). The plasma produced inside MC is uniform over a diameter of more than 4 cm at the center and ~ 2 cm at the meniscus (edge of the MC) from where the beams are extracted, therefore it is expected that the beam will be uniform through all seven apertures of PE. It may be noted that in the case of filament-generated plasmas, the radial uniformity often has to be compromised and this could lead to beam current nonuniformity on the substrate, where modification by the ion beam is planned. Differential pumping using a second turbo molecular pump in the beam extraction side is employed to maintain a good vacuum in the extraction side, and to avoid collision of the beam ions with residual neutral atoms or molecules. Without these arrangements the evacuation of the extraction chamber was

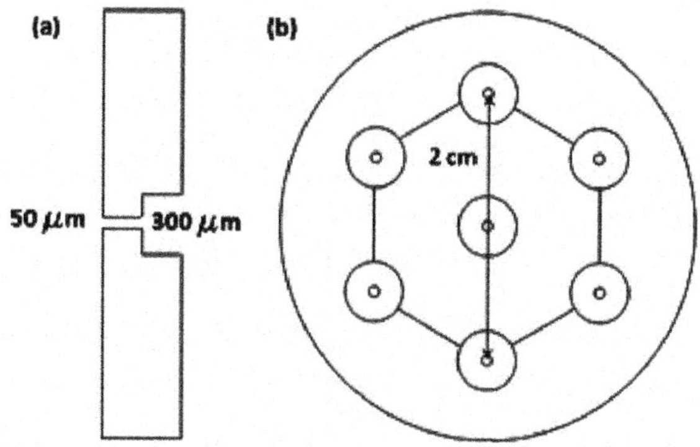

Fig. 3.2 The design of the PE with slotted aperture **a** cross-sectional view for one of the apertures and **b** front view (reprinted from [Abhishek Chowdhury and Sudeep Bhattacharjee, Journal of Applied Physics **107**, 093307 (2010)], with the permission of AIP Publishing)

only possible through the small holes of the plasma electrode. Differential pumping at the extraction chamber helped us to realize a vacuum better than ~10^{-5} Torr (at the ion extraction side) even when the plasma chamber was maintained at ~10^{-4} Torr.

3.3 Design of Plasma Electrodes for Generation of Broad and Micro-ion Beamlets

In most experimental setups, two or three distinct plasma electrode (PE) designs were employed based on specific experimental requirements. These designs include

(i) Simple through aperture: This configuration consists of a straightforward circular hole, the size of which can be adjusted depending on the need for broad ion beams. It provides uniform ion beam extraction but lacks significant focusing capability.
(ii) Slotted through aperture: In this design, a slot is incorporated within the aperture, allowing for controlled beam shaping. When multiple such slots are implemented, they facilitate the generation of micro-ion beamlets, offering precise ion beam localization.

(a) Focusing Effect of Slotted Aperture and Beam Optimization

The focusing behavior of these apertures was systematically analyzed, and it was observed that the slotted aperture design (Fig. 3.2a) exhibited superior focusing capability. This enhancement allows for the generation of well-defined micro-ion beamlets, making it a preferred choice when focused beams are required. To optimize the ratio between the through-aperture diameter and the slot width, AXCEL-INP simulation code [26] was employed. The simulations revealed that an optimal diameter-to-slot ratio of 1:6 yielded the best beam-focusing effect. This ratio ensures that the extracted ions are efficiently collimated while maintaining adequate beam intensity. For the extraction of collimated beam with adequate beam intensity this ratio is to be maintained. The enhancement in beam focusing can be assured with this simple through aperture. When multiple slots are incorporated, this configuration effectively facilitates the generation of micro-ion beamlets. The plasma electrode (PE) is in direct contact with the plasma, where the plasma potential typically falls within the range of 50–80 V. For a single through-aperture configuration aimed at generating broad ion beams, both the aperture diameter in the PE and the electrode thickness are maintained at 1 mm. The separation distance (d) between the PE and the collector (C) is fixed at 4 mm, while the applied collector potential (V) can be adjusted as needed.

The ion beam fluence, defined as the number of ions incident per unit area extracted from the plasma source, is tunable within the range of approximately 4.4×10^{15}–7×10^{15} ions cm^{-2}. Broad ion beams, with a diameter of approximately 20 mm, are employed to ensure uniform irradiation of target samples measuring 10 mm × 10 mm, with thicknesses ranging from 80 to 200 nm. In specific cases, thicker samples were also utilized to investigate the effects of ion irradiation on bulk-like material properties.

An alternative design was fabricated for further reduction of the beam spot size which is a plasma electrode (PE) with a smaller aperture size. A 5 × 5 array of holes was created on a 0.5-mm-thick stainless-steel plate, as illustrated in Fig. 3.3. These apertures consist of simple through holes, each with a diameter of 100 μm, produced using laser drilling. Due to the current limitations of the available laser system, 100 μm represents the smallest achievable aperture size. The spacing between the centers of adjacent apertures is maintained at 1 mm, as shown in Fig. 3.3a. A digital photograph of the fabricated PE is presented in Fig. 3.3b.

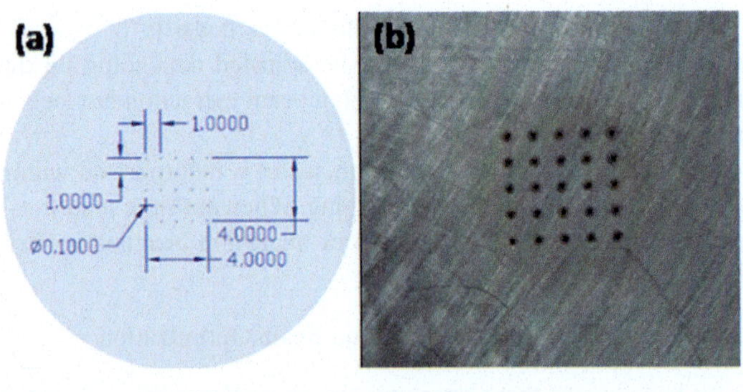

Fig. 3.3 **a** The design and **b** photograph of plasma electrode with aperture array pattern having 100-micron holes (reprinted from [Abhishek Chowdhury and Sudeep Bhattacharjee, AIP Advances 1, 042150 (2011)], with the permission of AIP Publishing)

3.4 AXCEL-INP and SIMION Simulations

Just as optical lenses that focus light, ion beams can be focused using electrostatic lenses, where the shape of the equipotential lines plays a crucial role in beam convergence. A convex equipotential surface facing the incoming ion beam exerts a radial inward force, effectively focusing the beam. Therefore, the design of focusing electrodes must ensure the generation of such convex equipotential lines to achieve proper beam focusing. To optimize electrode design, numerical simulations are typically performed prior to fabrication using established beam simulation tools such as AXCEL-INP [26] and SIMION [27]. These simulations allow for iterative adjustments by applying different trial potentials to the electrodes and analyzing the resulting equipotential field profiles. AXCEL-INP, a Vlasov solver based on the finite difference method, simultaneously solves the Poisson equation and the particle distribution function. The simulation begins with initial input conditions to generate a two-dimensional potential map, from which particle trajectories are calculated. A space charge map is then generated and used iteratively to refine the potential distribution and ion trajectories until a self-consistent solution is obtained. Essential input parameters for AXCEL-INP simulations include ion mass and charge, ion current density, plasma potential, and electron and ion temperatures. On the other hand, SIMION employs a modified fourth-order Runge-Kutta method to solve Laplace's equation and compute electrostatic or magnetostatics potential fields. While it does not directly solve the Poisson equation, it incorporates charge repulsion methods to approximate space charge effects and particle interactions. Ion trajectories in SIMION are determined by the combined influence of electrostatic forces and space charge repulsion, based on their instantaneous position and velocity. Key input parameters for SIMION simulations include ion mass and charge, initial kinetic energy, ion source position, and the total number of ions under consideration (Fig. 3.4).

3.5 Beam Current, Space Charge Effects, and Beam Extraction

To gain deeper insights into beam current dynamics and ion extraction mechanisms, including space charge effects, a comprehensive series of experiments was conducted using the specially designed plasma electrodes (PE) described earlier. These investigations aimed to optimize ion beam extraction efficiency and assess the impact of electrode geometry on beam characteristics. For initial validation of the PE design, the total extracted beam current was systematically measured while varying the extraction voltage across a wide range (0.01–3.0 kV). The plasma electrode consists

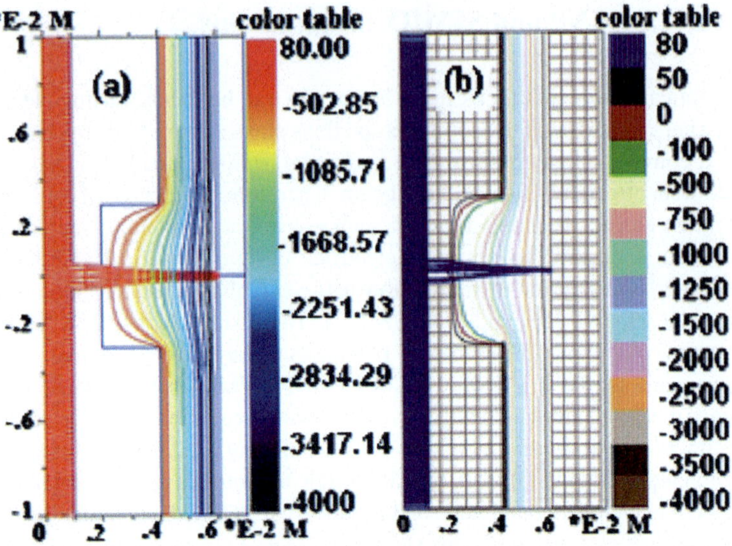

Fig. 3.4 Focused ion beam trajectories and equipotential lines as obtained in **a** AXCEL-INP and **b** SIMION simulations, with −4 kV collector potential for slotted aperture (units are in $\times 10^{-2}$ m) (reprinted from [Abhishek Chowdhury and Sudeep Bhattacharjee, Journal of Applied Physics **107**, 093307 (2010)], with the permission of AIP Publishing)

of two distinct surfaces: the plasma-facing side, which features a continuous metallic layer with precisely engineered apertures and is maintained at ground potential, and the rear side, which houses the electrical connection pads for individual beamlet control. This configuration enables fine-tuned modulation of the beamlet properties, facilitating precise manipulation of the extracted ion flux for various experimental applications.

To enable precise control over individual beamlets, a custom-designed switching circuit was developed to regulate the voltages applied to the individual pads of the plasma electrode (PE), as illustrated in Fig. 3.5b. This circuit facilitates independent beamlet modulation, allowing for highly controlled ion beam extraction and patterning. The circuit architecture integrates a 40-pin ATMEGA-16 L microcontroller unit (MCU), relays (HJR-3FF-S-Z), MOSFETs (FB A14 IRF830B), and precision resistances. The microcontroller serves as the central processing unit, orchestrating the switching operations required for dynamic voltage control. The relays and MOSFETs function as high-speed electronic switches, enabling real-time modulation of individual beamlets.

The relation between total beam current and extraction voltage is presented in Fig. 3.5a for plasma electrode (PE) apertures of 800 μm, 400 μm, and 200 μm. Initially, the beam current exhibits a rapid increase up to approximately 100 V, after

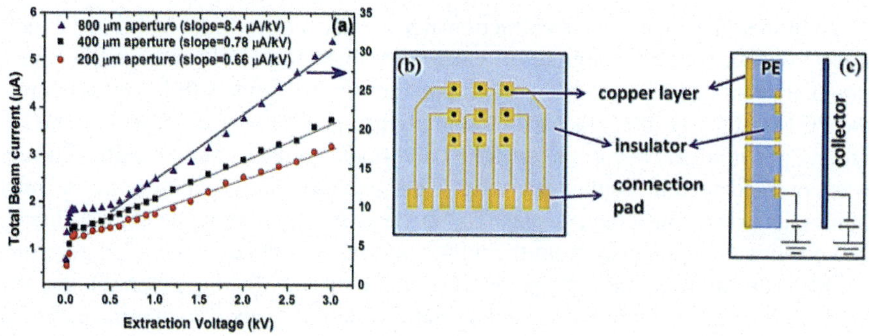

Fig. 3.5 **a** Variation of the total beam current with extraction voltage for three apertures of 800 μm, 400 μm, and 200 μm. **b** Schematic of the front view of PE, and **c** cross-sectional view of PE and collector with potentials (AIP open access article [28] distributed under the terms of the Creative Commons CC BY license)

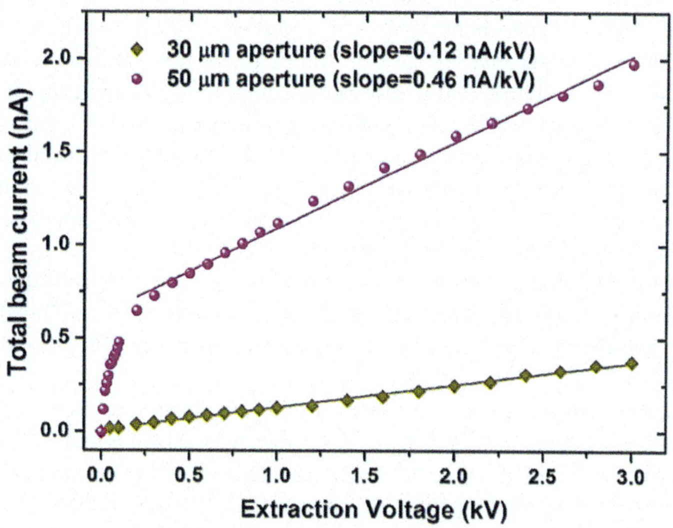

Fig. 3.6 Variation of the total beam current with extraction potential for 30 and 50 μm apertures (AIP open access article [28] distributed under the terms of the Creative Commons CC BY license)

which it increases more steadily with further increases in extraction voltage. In this steady-growth region, the rate of current increase, represented by the slope of the I_b versus V_E curve, shows a clear dependence on the aperture diameter. Specifically, larger apertures result in a steeper slope, with values ranging from 8.4 μA/kV to 0.66 μA/kV, as depicted in the figure. Notably, up to −3 kV, no saturation of the beam current is observed. The maximum ion current reached 32 μA at an extraction voltage of −3 kV for the 800 μm PE aperture, indicating robust current generation capabilities.

To further investigate beam extraction at reduced aperture sizes, additional experiments were conducted using plasma electrodes (PE) with significantly smaller aperture diameters of 50 μm and 30 μm. In previous experiments with larger apertures, the PE was directly in contact with the multicusp plasma source and maintained at ground potential. However, under these conditions, no measurable beamlet current was observed for the smaller apertures. This suggests that grounding the PE may inhibit ion extraction at these scales. To overcome this limitation, the PE was floated and positioned 4 mm away from the multicusp plasma source. In this configuration, the PE naturally acquired a potential equal to the plasma floating potential, thereby enabling beamlet extraction even at the smallest aperture sizes. The floating potential acquired by an electrode immersed in a plasma is given by [29].

$$V_f = -\frac{\kappa_B T_e}{e} \ln\left(\frac{m_i}{4\pi m_e}\right), \tag{3.1}$$

where T_e is the electron temperature and m_i and m_e are ion and electron masses, respectively. In our experiments with Argon plasma, T_e at the extraction region is ~5 eV and V_f is obtained to be ~ − 21.7 V. Therefore, the positive ions are accelerated through this extra potential and could pass through the smaller PE apertures. The extracted current obtained in this condition is plotted in Fig. 3.6. Similar trend of current is observed, although the magnitude of the current is reduced quite a bit for these smaller apertures as shown in Fig. 3.6. The slope of I_b–V_E plot is much smaller as expected for these smaller apertures and increases with aperture diameter in agreement with the larger aperture size PEs.

Except for the reduced aperture diameter, another possible reason for the reduction in the current is that the plasma electrode is kept 4 mm away from the edge of the multicusp and the ion density as well as ion saturation current falls sharply outside the multicusp.

Plasma electrodes (PE) featuring an array of through apertures are widely utilized for ion beam extraction. In cases where highly focused ion beams with minimal beam divergence are required, the aperture diameters are designed to be comparable to the plasma Debye length, enabling the generation of ultra-fine ion beamlets. Notably, ion extraction through such sub-Debye length apertures is feasible when the PE is maintained at floating potential. Experimental observations reveal that the current-voltage (I–V) characteristics of the extracted beamlets exhibit unconventional behavior. Specifically, the space-charge-limited region deviates from the predictions of the conventional Child-Langmuir law, suggesting modified space-charge effects at these small scales. Additionally, the extraction-voltage-limited regime does not exhibit conventional saturation but instead follows a Schottky-like behavior akin to that observed in vacuum diodes. These findings provide new insights into ion beam extraction physics and have significant implications for precision ion beam applications in nanofabrication and surface engineering.

When a potential of lower magnitude is applied on the collector, the beam current reaching the collector remains lower than the total current available for extraction. However, ion beam extraction persists even at these small voltages, albeit with significantly reduced currents. This leads to the formation of an ion cloud near the plasma electrode (PE), where space charge effects dominate and regulate the extracted current. The repulsive influence of the space charge imposes a fundamental constraint on the number of ions that can emerge from the apertures, effectively limiting the beam current. Conventionally, in such space-charge-limited conditions, the current–voltage (I–V) relationship is expected to follow the Child-Langmuir law, which predicts a dependence of ($I \propto V^{3/2}$). However, this classical model is derived under idealized assumptions, including infinite planar electrodes, parallel equipotential surfaces, and zero initial ion velocity, none of which are fully satisfied in the experimental setup. Notably, when the aperture size approaches the Debye length λ_D, a significant deviation from the traditional Child-Langmuir scaling is observed. Experimental results reveal an alternative I–V dependence given by $I = 1.82\ V^{1/2}$, which diverges from the classical $V^{3/2}$ behavior. This empirical relationship has been plotted against experimental data in Fig. 3.7, demonstrating reasonable agreement. A detailed derivation of this modified scaling law is provided in Appendix 1. These findings suggest a fundamental shift in space-charge-limited transport dynamics at sub-Debye length scales, with important implications for precision ion beam extraction and nanoscale charge transport phenomena.

As illustrated in Fig. 3.6, beyond the space-charge-limited flow regime, the ion current exhibits a tendency toward saturation. This saturation corresponds to a state where all ions emitted through the plasma electrode (PE) apertures successfully

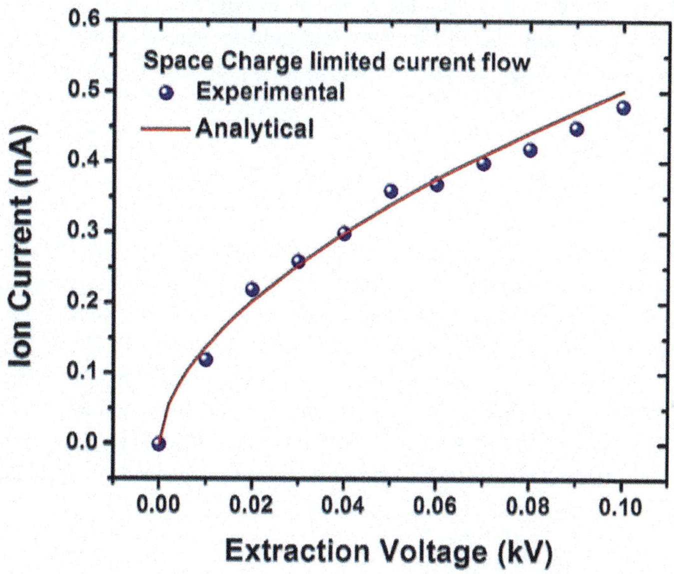

Fig. 3.7 A plot to compare the experimental data in the space-charge-limited flow region with the equation $I = 1.82\ V^{1/2}$ (AIP open access article [28] distributed under the terms of the Creative Commons CC BY license).

reach the collector. In an ideal scenario, once the system surpasses the space-charge-limited region, the total ion current should remain independent of the collector potential. However, experimental observations reveal a deviation from this theoretical expectation. Notably, Fig. 3.6 demonstrates a gradual increase in ion current with rising collector potential, a behavior reminiscent of the well-known Schottky effect observed in vacuum diode valves. In the classical Schottky effect, an increase in collector potential reduces the work function of the emitting surface, thereby enhancing electron emission. A similar mechanism could be at play in our experimental system, where the collector potential influences the potential distribution near the PE apertures, likely due to the formation and dynamics of the plasma sheath at the PE boundary.

To gain deeper insight into this phenomenon, we investigate the modification of the potential distribution at the aperture center in response to the applied collector potential. Figure 3.8 presents the simulated ion beam trajectories and corresponding equipotential contours, obtained using the AXCEL-INP simulation code. The results reveal that the equipotential surfaces, shaped by the collector potential, extend through the PE apertures and penetrate into the plasma sheath region. These curved equipotential lines effectively function as electrostatic lenses, focusing the ion beam onto the collector, as depicted in Fig. 3.8. This interplay between collector potential and beam trajectory highlights an essential mechanism governing ion beam extraction and focusing, with potential implications for advanced ion beam applications.

The part that has been simulated (cf. Fig. 3.8) is therefore the plasma region and the region between the PE and C. The figure shows the emergence of the beam from the plasma side through the PE aperture and getting focused onto the collector C. For this kind of circular aperture of diameter, a, the current density is calculated to be

$$J_{\mathrm{i}}(\eta_{\mathrm{w}}) = J_{\mathrm{io}} \exp\left[\frac{e}{\kappa T_{\mathrm{e}}}\left(\frac{\varphi_{\mathrm{o}}}{z_{\mathrm{o}}} - \frac{\varphi_1}{z_1}\right)\frac{a}{\pi}\right], \tag{3.2}$$

where φ_{o} is the collector potential, $z_{\mathrm{o}} = 2$ mm, the distance between PE and collector C, $\mathrm{E}_1 = \frac{\varphi_1}{z_1}$, φ_1 is the potential at the center of the aperture with respect to the bulk plasma potential, and z_1 may be taken as equal to the sheath thickness ($= n \times$ Debye length) (we have taken $n = 2$ for the calculations). Equation (3.2) has been plotted with the experimental data and shown in Fig. 3.9 and agrees with the experimental I–V relationship within the extraction-voltage-limited flow regime. Thus, it gives a more or less satisfactory explanation of the observed Schottky-like effect for our experimental system. The steps to derive Eq. (3.2) are outlined in Appendix 2.

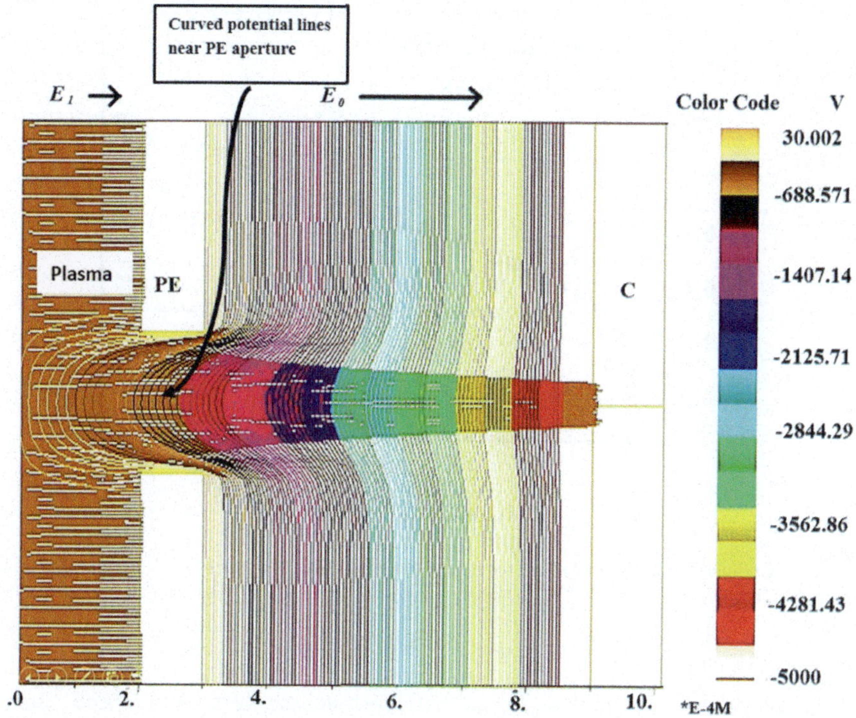

Fig. 3.8 AXCEL-INP simulation showing the ion beam trajectory and the equipotential surfaces (AIP open access article [28] distributed under the terms of the Creative Commons CC BY license)

3.6 Measurement of the Total Beam Current

3.6.1 Honeycomb-Patterned PE

From measurements of the ion saturation current, a working pressure of 0.25 mTorr is chosen, as this provides a maximum value for the current and all subsequent beam extraction experiments are performed at this pressure.

Ion beams are extracted by applying a negative bias of up to −5 kV to the collector while keeping the plasma electrode (PE) and the multicusp (MC) chamber at ground potential. This high voltage serves the dual purpose of extracting and focusing the ion beam, influenced by the nature of the equipotential lines, as illustrated in Fig. 3.8. The ion beam current I_b at the collector is measured both for individual apertures and for the total beam current through all apertures of the honeycomb-patterned PE (see Fig. 3.2). To isolate the current from a single aperture, the remaining apertures are covered with aluminum foil. The extraction potential is systematically varied from 0 to −5 kV, and the corresponding I_b values are recorded at the collector plate. Throughout these measurements, the microwave power is maintained at 270 W to

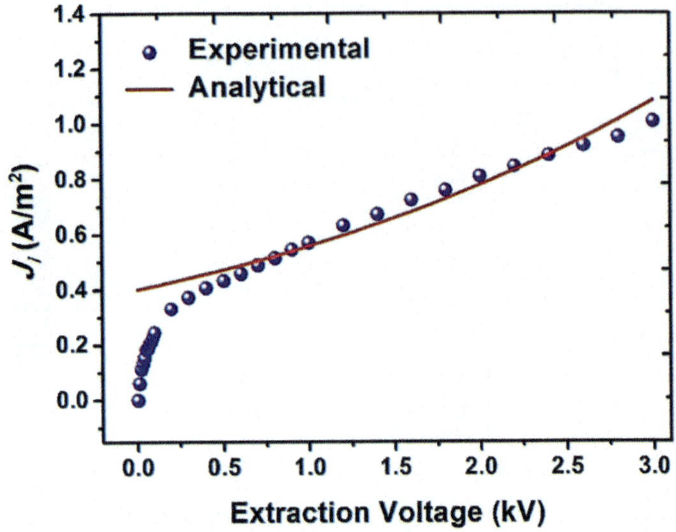

Fig. 3.9 Comparison of Eq. (3.2) with the experimental results for extraction-voltage-limited flow regime (AIP open access article [28] distributed under the terms of the Creative Commons CC BY license)

ensure stable plasma conditions. Figure 3.10 illustrates the variation of I_b with extraction voltage. A noticeable transition occurs at approximately 100 V, as shown in the inset of Fig. 3.10. In our experimental setup, the plasma potential V_p is in the range of 30–40 V. Consequently, the energy of the extracted ion beams corresponds to this potential. When the collector potential is lower than V_p, the ions experience a decelerating field, leading to beam defocusing and reduced collection efficiency. As the extraction voltage increases beyond V_p, a greater number of ions successfully reach the collector, culminating in a transition voltage where nearly all extracted ions are collected. Beyond this threshold, the ion current exhibits a gradual saturation trend until the extraction limit is reached. Experimental results indicate that individual beamlet currents peak at approximately 7 μA, while the total extracted beam current reaches ~51 μA. The total current exceeds seven times the individual aperture currents, suggesting potential secondary effects such as beam overlap or enhanced extraction efficiency at higher voltages. These findings offer valuable insights into optimizing focused ion beam generation and improving beam extraction efficiency.

3.6.2 Aperture Array-Patterned PE

From the above experiment, it is seen that the extracted current through the apertures is more or less uniform.

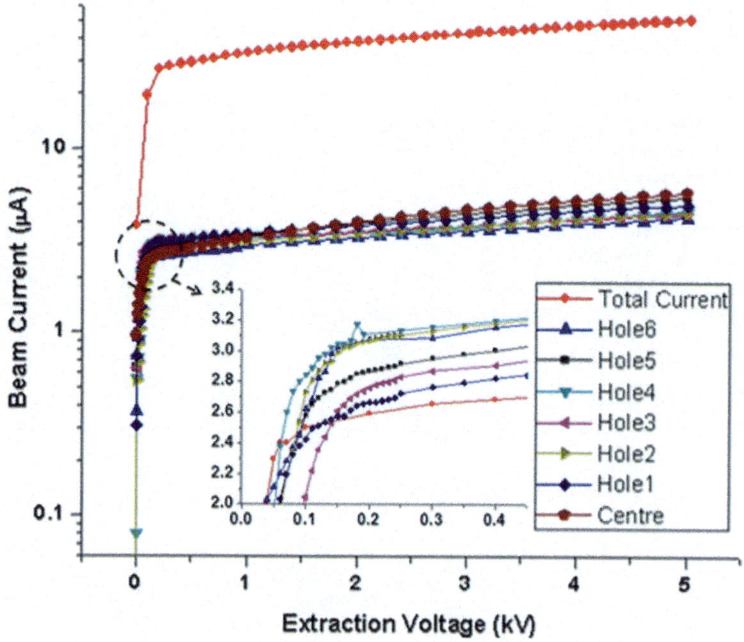

Fig. 3.10 Variation of beam current with extraction potential for honeycomb-patterned plasma electrode (inset: magnified view of transition region) (reprinted from [Abhishek Chowdhury and Sudeep Bhattacharjee, AIP Advances 1, 042150 (2011)], with the permission of AIP Publishing)

In this experiment, the total extracted ion beam current I_b is measured at the collector for all 25 apertures. The extraction voltage is systematically varied from 0 to −3 kV, and the corresponding I_b values are recorded. The gas pressure and microwave power are maintained identical to the conditions used in the previous experiment with the honeycomb-patterned plasma electrode (PE). Figure 3.11a presents the variation of I_b as a function of extraction potential for different gas pressures while keeping the microwave power constant. In contrast, Fig. 3.11b illustrates the dependence of I_b on microwave power for a fixed pressure. The inset in Fig. 3.11b provides a magnified view of the initial transition region. The results reveal that I_b increases with extraction voltage and does not exhibit saturation within the explored voltage range. A comparison of the extracted current-voltage characteristics for the two configurations (Figs. 3.10, 3.11) indicates that the honeycomb-patterned PE yields a higher total beam current due to its larger effective aperture area. However, beyond 250 V, the extracted current appears to be less sensitive to further increase in voltage. The differences in current-voltage characteristics between the two configurations can be attributed to the distinct electric field profiles at the apertures, specifically, the difference between a notched versus a simple through-aperture geometry. These findings provide critical insights into optimizing aperture design for efficient ion beam extraction.

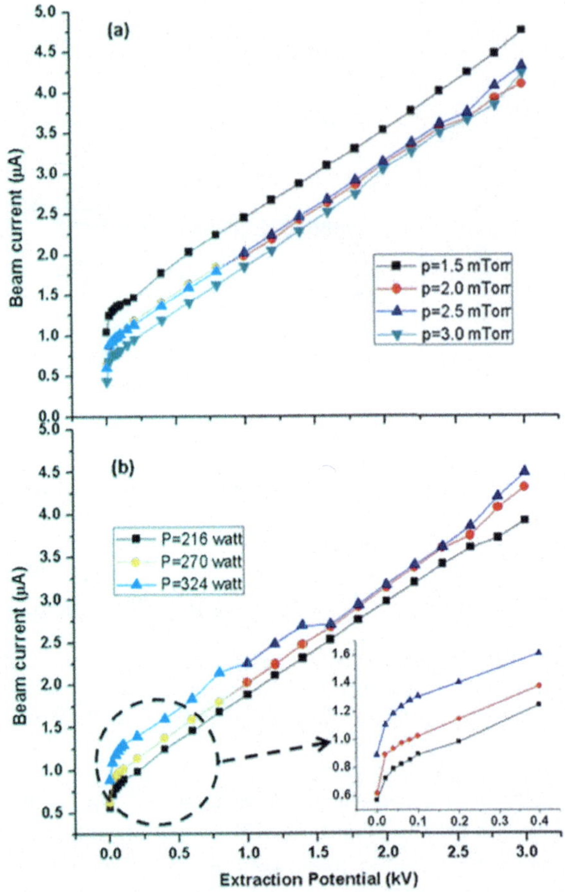

Fig. 3.11 Variation of beam current with extraction potential for **a** different plasma pressure keeping power = 270 W and **b** different microwave power keeping pressure = 0.25 mtorr (inset: magnified view of transition region) (reprinted from [Abhishek Chowdhury and Sudeep Bhattacharjee, AIP Advances 1, 042150 (2011)], with the permission of AIP Publishing)

3.7 Measurement Methods of Beam Uniformity

3.7.1 *Testing Beamlet Current Uniformity from Multiple Micro-apertures on Substrates*

After the determination of the stopping potential V_S, which is the minimum voltage that need to be applied to the selected apertures for stopping extraction of the beam, the individual beamlet current from all the nine apertures for the PE configuration of Fig. 3.5b was measured separately using a controller circuit. The details of the measurement of the stopping potential and the controller circuit are presented in the next chapter. In the nine-aperture (3 × 3) system, the apertures could be selectively switched on and off. The results are shown in Fig. 3.12.

Figure 3.12a shows the individual currents for the 200 μm aperture, whereas in Fig. 3.12b, the results for the 50 μm aperture are shown. In both the cases, the

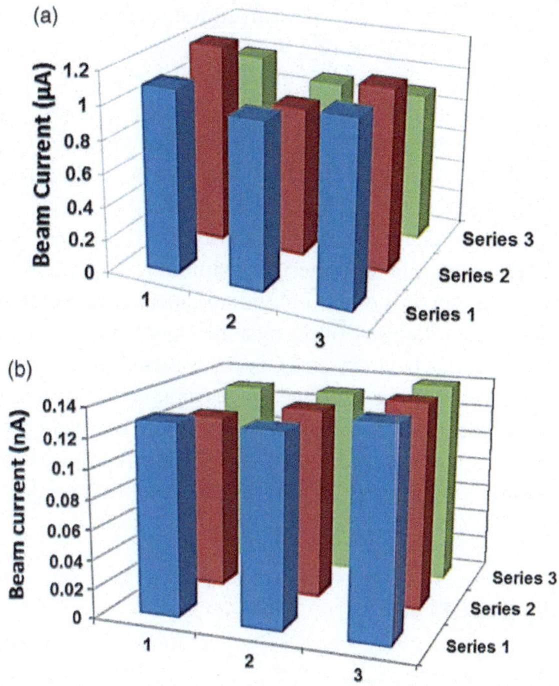

Fig. 3.12 Individual beamlet currents from 9 apertures for the PE with aperture **a** 200 μm and **b** 50 μm (AIP open access article [28] distributed under the terms of the Creative Commons CC BY license)

extraction potential is kept fixed at −1 kV. There is a small variation in currents for the 200 μm apertures, but it is almost uniform for the 50 μm apertures. The reason for this slight non-uniform distribution of beam currents in (3 × 3) array is that the apertures are not identical to each other. There are small variations in geometrical parameters (i.e., diameter and roundness) of the apertures. Apertures of 200 μm and larger were drilled mechanically and thus more variations in shape and size are observed. Whereas the smaller apertures are laser drilled and provide more accurate and almost identical apertures and therefore more uniform beamlet current through all the apertures.

3.8 Conclusion and Summary

The study demonstrates the Multi-Ion Beam System (MIBS) [31, 32] as a versatile tool for ion beam patterning via a beam-switching technique. Individual beamlet control is achieved by applying a small repulsive potential (0–120 V) to selected apertures, enabling dynamic patterning with applications in surface modification and material processing. The extracted beam current profile reveals two distinct operational regimes: (a) a space-charge-limited flow region, where the current-voltage relation deviates from the conventional Child-Langmuir law and (b) an extraction-voltage-limited flow region, where the beam current increases gradually instead of saturating, resembling the Schottky effect in vacuum diodes. This behavior is attributed to the penetration of collector-induced equipotential surfaces into the plasma sheath. As the aperture size decreases below the Debye length, the influence of the collector potential diminishes, leading to an increment in the extraction of the beam governed by the sheath potential developed at the center of the aperture. With the integration of a switching electrode, the system mimics a vacuum triode valve, exhibiting characteristics similar to its mutual or transfer function. Controlled beamlet extraction and individual current measurements confirm uniformity, validating the reliability of the system. Furthermore, successful pattern formation using the different beamlet array is demonstrated. The irradiated spots act as an active center for advanced surface processing, including chemical etching, electrochemical treatments, selective material deposition, and self-organized microstructure formation. These findings establish the potential outlook for precision nanofabrication and engineering of material for scientific and industrial applications.

Appendix 1

In order to find out the current–voltage relationship for our system in the space-charge limited flow region, let us assume that the I–V relationship is given by

$$I = aV^{\gamma} + b. \tag{3.3}$$

Since, at $V = 0$, $I = 0$, we are left with $b = 0$. Taking natural log on both sides, we obtain

$$\ln(I) = \ln(a) + \gamma \ln(V). \tag{3.4}$$

Therefore, a plot between $\ln(I)$ and $\ln(V)$ will be a straight line whose slope is γ and intercept on the y-axis is $\ln(a)$.

From Fig. 3.13, it is found that the slope is 0.5 and intercept is $\ln(a) = 0.6$. Therefore, the actual I–V relationship that is found for our experiments is

$$I = 1.82V^{1/2} \tag{3.5}$$

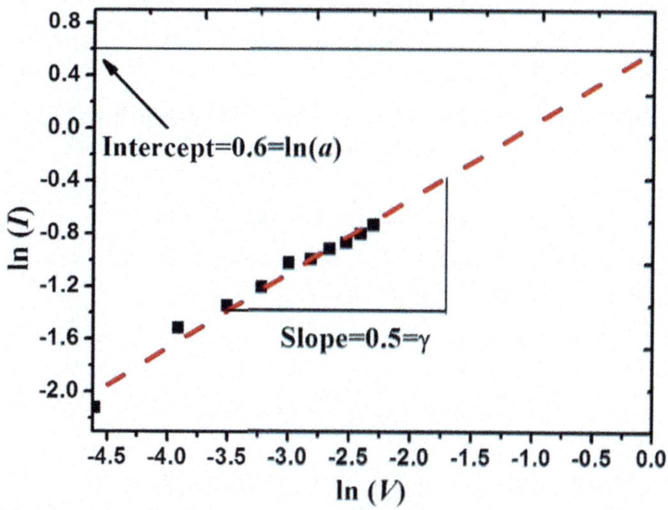

Fig. 3.13 ln (I) versus ln (V) in space charge limited flow region (AIP open access article [28] distributed under the terms of the Creative Commons CC BY license)

Appendix 2

Now, to calculate the I–V relationship in the extraction-voltage-limited region we take the following approach. The potential at the center of the circular aperture of radius a is given by [29].

$$\phi^{(1)} = (E_0 - E_1)\frac{a}{\pi}, \tag{3.6}$$

where E_0 and E_1 are the electric fields on both sides of the aperture—in this case, they are electric fields due to the collector and the electric field due to the formation of sheath, respectively, marked in Fig. 3.8.

The ion current density at any point inside the sheath is [30] given by

$$J_{i0}(\eta) = \left(\frac{2\kappa_B T_e}{M_i}\right)^{\frac{1}{2}} n_{e0} e I(\eta), \tag{3.7}$$

where T_e is the electron temperature, M_i is the ionic mass, and n_{e0} is the plasma electron density. $\eta = -e\phi/\kappa T_e$, where ϕ is the potential at any point inside the sheath with respect to the potential of the bulk plasma and $I(\eta)$ is a term which is a function of η and it needs to be determined. Assuming the electrons obey the Boltzmann distribution, the electron current at the PE wall is given by

$$J_e = n_e v e = n_{e0} \exp(-\eta) e \left(\frac{\kappa_B T_e}{2\pi m_e} \right)^{\frac{1}{2}}. \tag{3.8}$$

At equilibrium, the ion and the electron current will be equal at the PE wall. This gives

$$\eta_w = -\frac{e\phi_w}{\kappa_B T_e} = \ln\left[\frac{1}{2\sqrt{\pi}} \left(\frac{M_i}{m_e} \right)^{\frac{1}{2}} / I(\eta_w) \right], \tag{3.9}$$

which implies

$$I(\eta_w) = \frac{1}{2\sqrt{\pi}} \left(\frac{M_i}{m_e} \right)^{\frac{1}{2}} \exp(-\eta_w). \tag{3.10}$$

Therefore, substituting for $I(\eta_w)$ in Eq. (3.7), we obtain

$$J_{i0}(\eta_w) = \left(\frac{2\kappa_B T_e}{M_i} \right)^{\frac{1}{2}} n_{e0} e \frac{1}{2\sqrt{\pi}} \left(\frac{M_i}{m_e} \right)^{\frac{1}{2}} \exp\left(\frac{e\phi_w}{\kappa_B T_e} \right) \tag{3.11}$$

where ϕ_w is the value of potential ϕ at the PE wall. In the presence of the collector potential, ϕ_w should be replaced by $(\phi_w + \phi^{(1)})$, where $\phi^{(1)}$ is given by Eq. (3.6). Defining the constant pre-exponential factor of Eq. (3.11) by A, we can write the current in the presence of the collector potential as

$$J_i(\eta_w) = A \exp\left(\frac{e}{\kappa_B T_e} (\phi_w + \phi^{(1)}) \right) = J_{i0}(\eta_w) \exp\left(\frac{e\phi^{(1)}}{\kappa_B T_e} \right). \tag{3.12}$$

By virtue of Eq. (3.6),

$$J_i(\eta_w) = J_{i0} \exp\left(\frac{e}{\kappa_B T_e} \left(\frac{\phi_0}{z_0} - \frac{\phi_1}{z_1} \right) \frac{a}{\pi} \right). \tag{3.13}$$

Questions

1. What are the potential applications of low-energy multiple ion beamlets?
2. What kind of energies do low-energy ion beams correspond to? What are the important trends in current research utilizing low-energy ion beams?
3. Please read the literature and find out how does a multicusp device provide plasma confinement in the radial and axial directions.

4. Please compare the working pressure of the neutral gas in the plasma chamber which is mentioned to be in the range 0.085 to 0.55 mTorr with atmospheric pressure. Familiarize yourself with the different units for measuring pressure such as Pascal and Torr. Although the Pascal is the SI unit for pressure, Torr is very commonly used while referring to neutral pressure in plasma systems.
5. What is meant by plasma potential? What potential does an electrode immersed in a plasma attain?
6. What are plasma sheaths? How thick is a plasma sheath in comparison with the Debye length of a plasma?
7. How does a plasma electrode with a notch have focusing effect on the ions that are extracted from the plasma?
8. What is the floating potential of an electrode immersed in an oxygen plasma having an electron temperature of 10 eV?
9. What is Child-Langmuir's law? How does the extracted ion current through sub-Debye length plasma electrode apertures differ from the conventional Child-Langmuir's law?
10. What is the velocity with which ions enter the sheath edge (read about Bohm velocity)? If the plasma electrode is grounded what is the energy of the ions as they arrive at the PE aperture?
11. An electrode is immersed in a collision-less, Maxwellian plasma consisting of electrons and singly charged ions. The electrons have a temperature of $T_e =$ 3 eV, and the ion temperature is negligible ($T_i \approx 0$). The ion mass is $m_i = 40$ amu (assume argon plasma), and the plasma is quasi-neutral with equal electron and ion densities far from the electrode (n_0). Given that the charge of an electron $e = 1.6 \times 10^{-19}$ coulomb and the electronic mass $m_e = 9.1\times 10^{-31}$ kg, (a) Determine the floating potential V_f acquired by the electrode relative to the plasma potential. Use the Bohm criterion to estimate the ion velocity at the sheath edge and (b) Discuss how the floating potential would change if the electron temperature was increased to $T_e = 5$ eV.

References

1. W. Li, R. Minev, S. Dimov, G. Lalev, Patterning of amorphous and polycrystalline Ni78B14Si8 with a focused-ion-beam. Appl. Surf. Sci. **253**, 5404 (2007)
2. E. Platzgummer, H. Loeschner, G. Gross, Projection maskless patterning for nanotechnology applications. J. Vac. Sci. Technol. B **26**, 2059 (2008)
3. Y. Lee, W.A. Barletta, K.N. Leung, V.V. Ngo, P. Scott, M. Wilcox, N. Zahir, Multi-aperture extraction system with micro-beamlet switching capability. Nucl. Instrum. Methods Phys. Res. A **474**, 86 (2001)
4. H.E.M. Peres, F.J. Ramirez-Fernandez, High resistivity silicon layers obtained by hydrogen ion implantation. Braz. J. Phys. **27A**, 237 (1997)
5. J.-H. Kim, B. Tsuchiya, S. Nagata, T. Shikama, Effects of ion beam surface modification on proton conductivity of $BaCe_{0.9}Y_{0.1}O_3$—δ. Solid State Ionics **179**, 1182 (2008)

6. E.R. Leber, B.D. Ratner, Introduction of carboxyl functional groups onto platinum by RF plasma deposition. Plasma Process. Polym. **6**, 219 (2009)
7. H.-J. Maas, J. Heimel, H. Fuchs, U.C. Fischer, J.C. Weeber, A. Dereux, Photonic nanopatterns of gold nanostructures indicate the excitation of surface plasmon modes of a wavelength of 50–100 nm by scanning near-field optical microscopy. J. Microsc. **209**, 241 (2003)
8. S. Macko, F. Frost, B. Ziberi, D. Forster, T. Michely, Is keV ion-induced pattern formation on Si(001) caused by metal impurities? Nanotechnology **21**, 085301 (2010)
9. B. Ziberi, F. Frost, T. Hoche, B. Rauschenbach, Ripple pattern formation on silicon surfaces by low-energy ion-beam erosion: experiment and theory. Phys. Rev. B **72**, 235310 (2005)
10. P.F.A. Alkemade, Propulsion of ripples on glass by ion bombardment. Phys. Rev. Lett. **96**, 107602 (2006)
11. H. Hansen, A. Redinger, S. Messlinger, G. Stoian, Y. Rosandi, H.M. Urbassek, Mechanisms of pattern formation in grazing-incidence ion bombardment of Pt(111). Phys. Rev. B **73**, 235414 (2006)
12. I. Koponen, M. Hautala, O.-P. Sievanen, Simulations of ripple formation on ion-bombarded solid surfaces. Phys. Rev. Lett. **78**, 2612 (1997)
13. S. Khakshouri, D.M. Duffy, Influence of electronic effects on the surface erosion of tungsten. Phys. Rev. B **80**, 085413 (2009)
14. R. Moroni, D. Sekiba, F. Buatier de Mongeot, G. Gonella, C. Boragno, L. Mattera, U. Valbusa, Uniaxial magnetic anisotropy in nanostructured Co/Cu (001): from surface ripples to nanowires. Phys. Rev. Lett. **91**, 167207 (2003)
15. P. Chaudhari et al., Atomic-beam alignment of inorganic materials for liquid-crystal displays. Nature **411**, 56 (2001)
16. L. Vattuone, U. Burghaus, L. Savio, M. Rocca, G. Costantini, F. Buatier de Mongeot, C. Boragno, S. Rusponi, U. Valbusa, Oxygen interaction with disordered and nanostructured Ag(001) surfaces. J. Chem. Phys. **115**, 3346 (2001)
17. S. Facsko, T. Dekorsy, C. Koerdt, C. Trappe, H. Kurz, A. Vogt, H.L. Hartnagel, Formation of ordered nanoscale semiconductor dots by ion sputtering. Science **285**, 1551 (1999)
18. R.S. Dhaka, C. Biswas, A.K. Shukla, S.R. Barman, Xe and Ar nanobubbles in Al studied by photoemission spectroscopy. Phys. Rev. B **77**, 104119 (2008)
19. P.B. Johnson, D.J. Mazey, Gas-bubble superlattice formation in bcc metals. J. Nucl. Mater. **218**, 273 (1995)
20. R.S. Dhaka, S.R. Barman, Plasmon excitations by photoelectron emission from rare gas Nanobubbles in aluminum. Phys. Rev. Lett. **104**, 036803 (2010)
21. C.J. Rossouw, S.E. Donnelly, Superheating of small solid-argon bubbles in aluminum. Phys. Rev. Lett. **55**, 2960 (1985)
22. M. Schmid, W. Hebenstreit, P. Varga, S. Crampin, Quantum Wells and electron interference phenomena in Al due to subsurface Noble gas bubbles. Phys. Rev. Lett. **76**, 2298 (1996)
23. D.S. Martin, R.J. Cole, P. Weightman, Effects of ion bombardment on the optical and electronic properties of Cu(110). Phys. Rev. B **72**, 035408 (2005)
24. M.P. Ray, R.E. Lake, C.E. Sosolik, L.B. Thomsen, G. Nielsen, I. Chorkendorff, O. Hansen, Subsurface excitations in a metal. Phys. Rev. B **80**, 161405. (R) (2009)
25. V.F. Reutov, A.S. Sokhatskii, Helium ion bombardment induced amorphization of silicon crystals. Tech. Phys. Lett. **28**, 615 (2002)
26. P. Spadtke, AXCEL-INP, Wiesbaden, Germany
27. D.A. Dahl, SIMION 7, Idaho National Engineering and environmental laboratory, Idaho Falls, ID. www.SIMION.com
28. A. Chowdhury, S. Chatterjee, A. Dutta, S. Bhattacharjee, Stopping potential and ion beamlet control for micro-resistive patterning through sub-Debye length plasma apertures. AIP Adv. **4**, 127127 (2014)
29. J.A. Bittencourt, *Fundamentals of Plasma Physics*, 3rd edn. (Springer, New York, 2004)
30. J.D. Jackson, Electrodynamics, 3rd ed (Wiley, 2011).

31. A. Chowdhury, S. Bhattacharjee, Localized subsurface modification of materials using micro-low-energy multiple ion beamlets. AIP Adv. **1**, 042150 (2011)
32. A. Chowdhury, S. Bhattacharjee, Micron-focused ion beamlets. J. Appl. Phys. **107**, 093307 (2010)

Chapter 4
Control of Micro-beamlets for Surface Patterning

The multi- and broad ion beam experimental setups were examined along with the characteristics of the extracted currents, as discussed in detail in the previous chapter. In order to process a large quantity of samples, the multiple ion beamlet system (MIBS) is a suitable alternative to the single ion beam irradiation system as discussed in Chap. 3. For large-scale surface structuring, the fabrication of functional surfaces on soft materials such as polymers [1], the design of intricate masks for micro-electronics, bio-patterning, and mask-less ion beam lithography [2, 3], the MIBS system is widely utilized. By employing multiple beamlets, pre-defined patterns can be generated over extensive sample areas, leading to improved cost-effectiveness and processing efficiency.

Several research groups around the world have reported the development of multi-ion beam systems (MIBS) for advanced lithography applications [2–4]. This includes: (a) filament-generated plasmas using Mask-less Micro-ion-beam Reduction Lithography (MMRL) developed by Leung et al. at the Lawrence Berkeley National Laboratory (LBNL) [3]. However, a key challenge with these plasma-based systems is achieving uniform radial plasma distribution, which can lead to non-uniform beam currents on the substrate [1], (b) a multi-ion beam lithography system reported by Appleton et al. at the University of Florida, capable of generating a 16×16 array of beamlets with energies ranging from 15 to 40 keV [5] and (c) an ion projection lithography (IPL) system developed in Europe, which uses a broad ion beam and a stencil mask to project patterns onto the sample [2].

The research outlined in Chap. 3 details the development of the MIBS in our laboratory to investigate the interaction of low-energy (1–5 keV) ion beams with various substrate materials at micrometer resolutions. The primary objective has been to create atomically heterogeneous systems, the specifics of which are provided in the next chapter, and to leverage these systems for diverse applications. The unique aspect of these systems is that the introduction of atomic-level heterogeneity generates new properties of material that are absent in parent material. Ongoing efforts aim to reduce the focused beamlet diameter to the sub-micron regime by optimizing the plasma

S. Bhattacharjee, *Atomically Heterogeneous Systems*,
https://doi.org/10.1007/978-981-95-9733-8_4

electrode (PE) aperture and improving beam focusing. This would enable the use of focused multiple ion beamlets for localized, patterned modification of substrates on the micron scale. Fabricating an array of small through holes in the PE, placed at the plasma source exit, allows for easier extraction of the ion beams from the plasma-facing electrode compared to a single through-hole with a notch, as the dimensions become increasingly diminutive [6, 7]. The size of the aperture can significantly impact the behavior of the plasma sheath and ion beam extraction, as discussed in Chap. 3. However, when the aperture size is smaller than the plasma Debye length (as defined in Chap. 1). The Debye length is dependent on the electron temperature (T_e) and electron density (n_e). Under typical experimental conditions at the ion beam extraction point ($T_e \sim 5$ eV, $n_e \sim 3 \times 10^{10}$/cc) the Debye length ($\lambda_D$) is approximately 100 μm.

When the aperture size is smaller than the λ_D, the extraction electric field and plasma sheath field at the aperture undergo significant, nonlinear changes [8, 9]. This affects the beam-focusing properties, leading to a notable enhancement of the electric field at the aperture. This in turn results in nonlinear demagnification and a reduction in the beam spot size [9]—an effect that cannot be achieved through conventional charged particle optics methods alone, such as simply applying electrostatic potentials to the extraction electrodes. The previous chapter discussed the extraction of multiple ion beamlets and the nature of the current-voltage (I–V) characteristics. This chapter will primarily focus on the design of a beam-switching circuit for beamlet control, the stopping potential of the beams (which enables patterning on substrates), and an example application of using the ion beamlets to modify the local resistivity of copper thin films in a patterned manner [7].

The experimental system described in Chap. 3 utilizes a purpose-built plasma electrode (PE) to control the individual ion beamlets. The PE features a three-layer design, with the first layer being a continuous metallic surface (except for nine apertures) facing the plasma, the second layer made of an insulator, and the third layer a patterned electrode capable of applying a voltage between 0 and 120 V. This voltage application creates a repulsive potential that can stop the beamlet current by applying a positive potential to the individual apertures of the PE. To extract the ion beamlets, a thin aluminum foil sample is kept 5 mm away from the PE on a collector (C) plate maintained at a high negative potential (1–3 kV). The layered PE design was fabricated in-house on a printed circuit board (PCB) using Protel software. PEs with varying aperture sizes (800, 400, and 200 μm) and the associated beam control system were produced on the PCB.

4.1 Design of Beam-Switching Circuits

A high-precision switching circuit has been designed and implemented to regulate the voltages on individual pads of the PE, enabling precise control over the individual beamlets. The schematic representation of the circuit is illustrated in Fig. 4.1a.

This system integrates a 40-pin ATMEGA-16L microcontroller unit (MCU), high-performance relays (HJR-3FF-S-Z), MOSFETs (FB A14 IRF830B), and carefully selected resistances to ensure optimal functionality. A digital photograph of the fabricated circuit is presented in Fig. 4.1b, demonstrating the practical realization of the design.

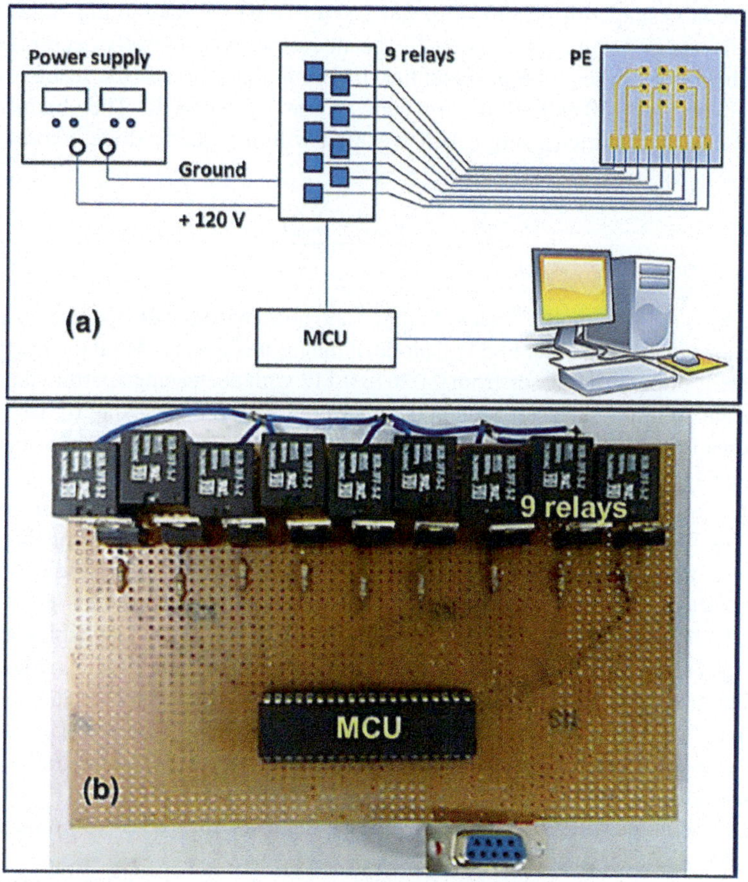

Fig. 4.1 **a** Schematic diagram of the beam-switching circuit and **b** digital photograph of the fabricated circuit consisting of a microcontroller unit (MCU), 9 relays, resistances and MOSFETs (AIP open access article [10] distributed under the terms of the Creative Commons CC BY license)

4.2 Stopping Potential and Control of Beam Extraction

The switching circuit is subsequently tested to determine the stopping potential voltages corresponding to each aperture diameter. This is achieved by measuring the total beam current at a fixed extraction potential while varying the voltage on the beamlet control pads, as illustrated in Fig. 3.5c of Chap. 3. The obtained results for the three largest apertures (800, 400, and 200 μm) are presented in Fig. 4.2a.

As the applied voltage increases, the beam current progressively decreases until it reaches zero at the stopping potential. Figure 4.3b illustrates the results for the smaller apertures (50 and 30 μm), revealing a comparable trend. The obtained data can be effectively fitted using a Hill equation with a variable slope, expressed as follows:

$$V = \frac{(y_{\max} - y_{\min})}{1 + 10^{(\log x_0 - x) \times p}}, \tag{4.1}$$

where $y_{\max}$ and $y_{\min}$ are the values at the top and bottom asymptotes, and $\log x_0$ and p are, respectively, the center and Hill slope of the fit. It is observed that the magnitude of the Hill slope, p, increases from 0.026 to 0.112 with decreasing aperture diameter, i.e., beam current falls faster for smaller apertures. This indicates that for larger size apertures the field penetration from the extraction side is more resulting in larger beam energy and beam current at the exit side of PE.

For a mono-energetic ion beam, the total beam current versus switching voltage curve (as depicted in Fig. 4.2) would ideally exhibit a step-function behavior. However, the experimental data follows the Hill equation (Eq. (4.2)), indicating that the extracted ions possess a distribution of energies centered around a mean value. By differentiating the total beam current (I_b) with respect to the switching voltage (V_s), the energy spread of the beam can be estimated, analogous to the method used in a retarding field energy analyzer. The data is well represented by a Gaussian fit, as shown in Fig. 4.3a and b for 200 μm and 30 μm apertures, respectively. The axial ion beam energy spread at the plasma electrode exit is determined from the full width at half maximum (FWHM) of the Gaussian distribution. As illustrated in Fig. 4.3, the FWHM increases with aperture diameter, ranging from 15.5 V for a 30 μm aperture to 21.5 V for a 200 μm aperture.

Figure 4.4 illustrates the dependence of (V_s) on the plasma electrode (PE) aperture size, revealing an approximately linear trend. A linear fit was applied to the data points, yielding an (R^2) value of 0.982, indicating a strong correlation. The numbers at each data point represent the standard deviations obtained from drilling nine apertures for each size (800, 400, 200, 50, and 30 μm). To validate the experimental results, comparisons were made with simulations performed using the standard beam optics code SIMION. In these simulations, the stopping potential was determined for each aperture size, and the corresponding results are represented by black squares in Fig. 4.4. While the simulation values are slightly lower than the experimental measurements, they exhibit the same linear trend. The observed discrepancy ranges from approximately 9% for the 800 μm aperture to about 30% for the 400 μm aperture.

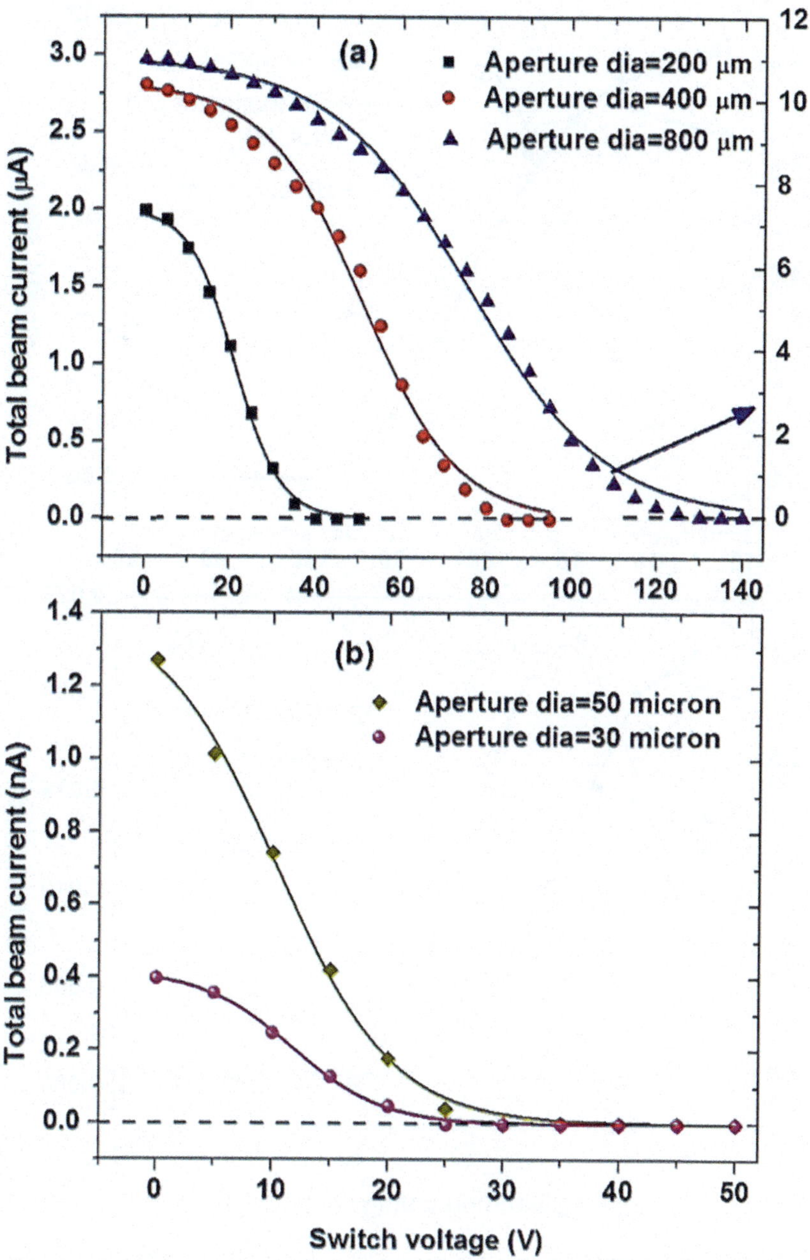

Fig. 4.2 The variation of the total beam current with the switching voltage **a** for bigger apertures (800, 400, and 200 μm) and **b** for smaller apertures (50 and 30 μm). Solid lines represent fitting according to Hill equation given by Eq. (4.1) (AIP open access article [10] distributed under the terms of the Creative Commons CC BY license)

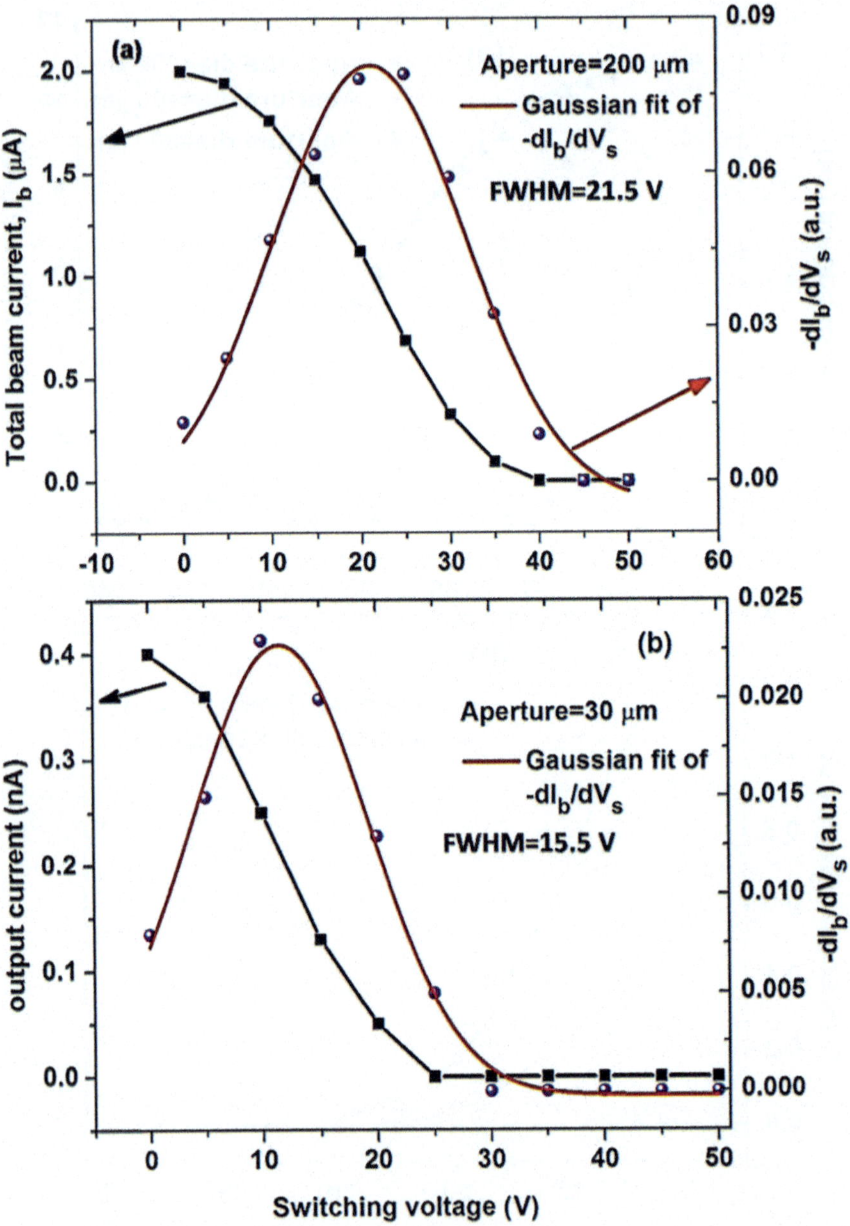

Fig. 4.3 Plot of total beam current and its derivative (blue circles) with respect to switch voltage. The derivative data is fitted with a Gaussian (red solid line) **a** for the 200 μm aperture and **b** for the 30 μm apertures (AIP open access article [10] distributed under the terms of the Creative Commons CC BY license)

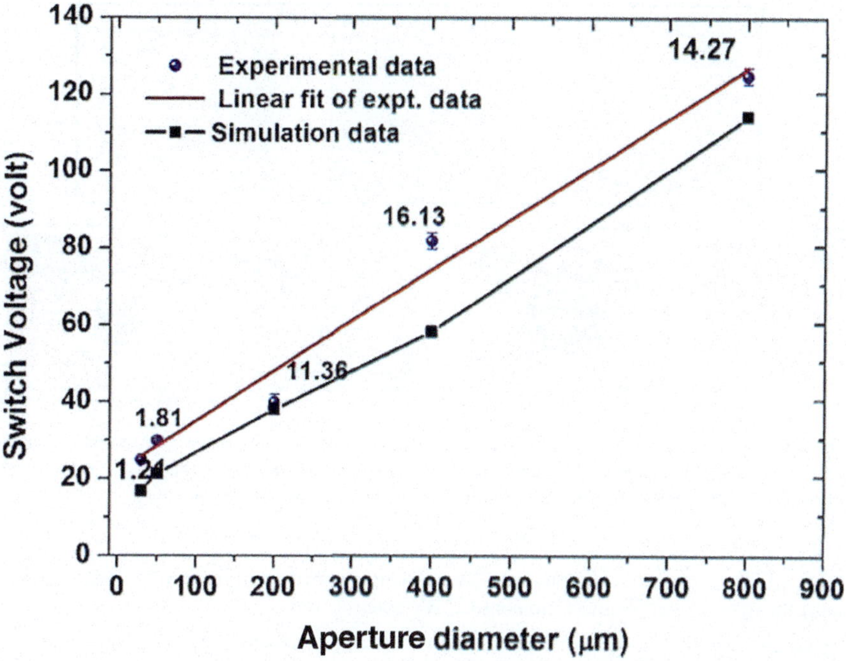

Fig. 4.4 Variation of the switch voltage V_0 with the aperture size. The numbers with the experimental data points indicate the standard deviation in diameter while drilling the nine apertures (AIP open access article [10] distributed under the terms of the Creative Commons CC BY license)

4.3 Generation of Surface Patterns

As discussed in Chap. 3, the individual beam currents measured from all nine apertures exhibited minor variations for the 200 μm aperture, while the 50 μm apertures produced more uniform results. Utilizing the switching circuit (depicted in Fig. 4.1), various patterns were successfully generated on substrates. To achieve this, a computer program was developed to control the microcontroller, enabling precise manipulation of the beamlets. The patterns were imprinted onto aluminum foils, and digital photographs of selected patterns are presented in Fig. 4.5. In Fig. 4.5a, all apertures remained open, while Fig. 4.5b-e displays distinct shapes: a square, a "U" shape, a "+" sign, and a right-angled triangle, respectively. Detailed specifications of these patterns are listed in Table 4.1.

The beam spot sizes at the substrate for a 50 μm PE aperture with extraction voltages of 2.5 kV and 2 kV are depicted in Fig. 4.5d and e, respectively. To validate some of the experimental results from ion beam irradiation, SIMION simulations were performed to determine the spot size at the focal point. In these simulations, the gap between the PE and the collector (where the substrate material is positioned) was kept constant, while the extraction potential was varied, and the corresponding beam diameter was measured. Changes in the extraction potential alter the electric

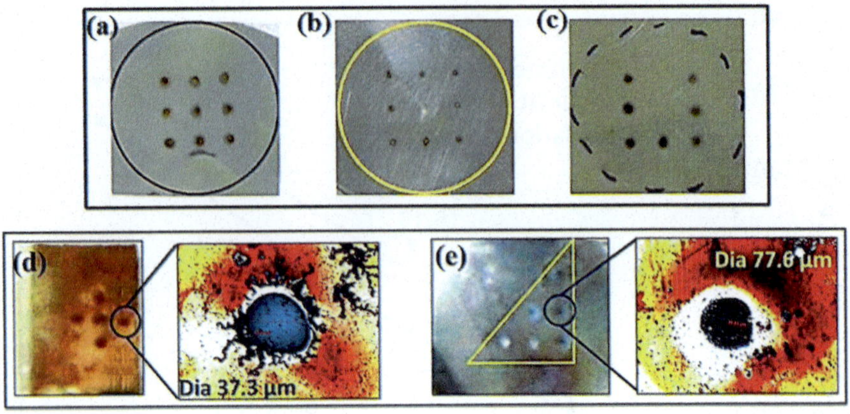

Fig. 4.5 The photographs of the patterned resistive region create in the aluminum foil **a** No patterning, **b** Square shape, **c** "U" shape, **d** a "+" shape, and **e** a "triangle" shape (AIP open access article [10] distributed under the terms of the Creative Commons CC BY license)

Table 4.1 Experimental conditions for generating patterns (AIP open access article [10] distributed under the terms of the Creative Commons CC BY license)

Serial no.	Shape	Aperture diameter (μm)	Extraction voltage (KV)	Irradiation time (s)
1	3 × 3	400	1	30
2	Square	200	0.5	20
3	"U"	800	1	60
4	"+"	50	2.5	120
5	Triangle	50	2.0	150

field between the PE and the collector, which in turn affects the focal point of the beam. The influence of the focal point on the beam by electric field was previously addressed by Chowdhury et al. [6]. In present experiments, it is aimed to achieve the smallest possible beam spot on the substrate, which was kept at a fixed distance from the PE. The results are presented in Fig. 4.6a for the three larger apertures (800, 400, and 200 μm), and in Fig. 4.6b for the two smaller apertures (50 and 30 μm). For the 50 μm aperture, the focal point was achieved at 0.3 kV, with a corresponding spot size of 6 μm. However, experimentally, a much larger spot size of 37.3 μm was observed (as shown in Fig. 4.5d). This discrepancy arises because, at 0.3 kV, no distinct spot formation on substrate is found. Consequently, the extraction voltage had to be increased, which shifted the focal point, resulting in a defocused larger spot on the substrate.

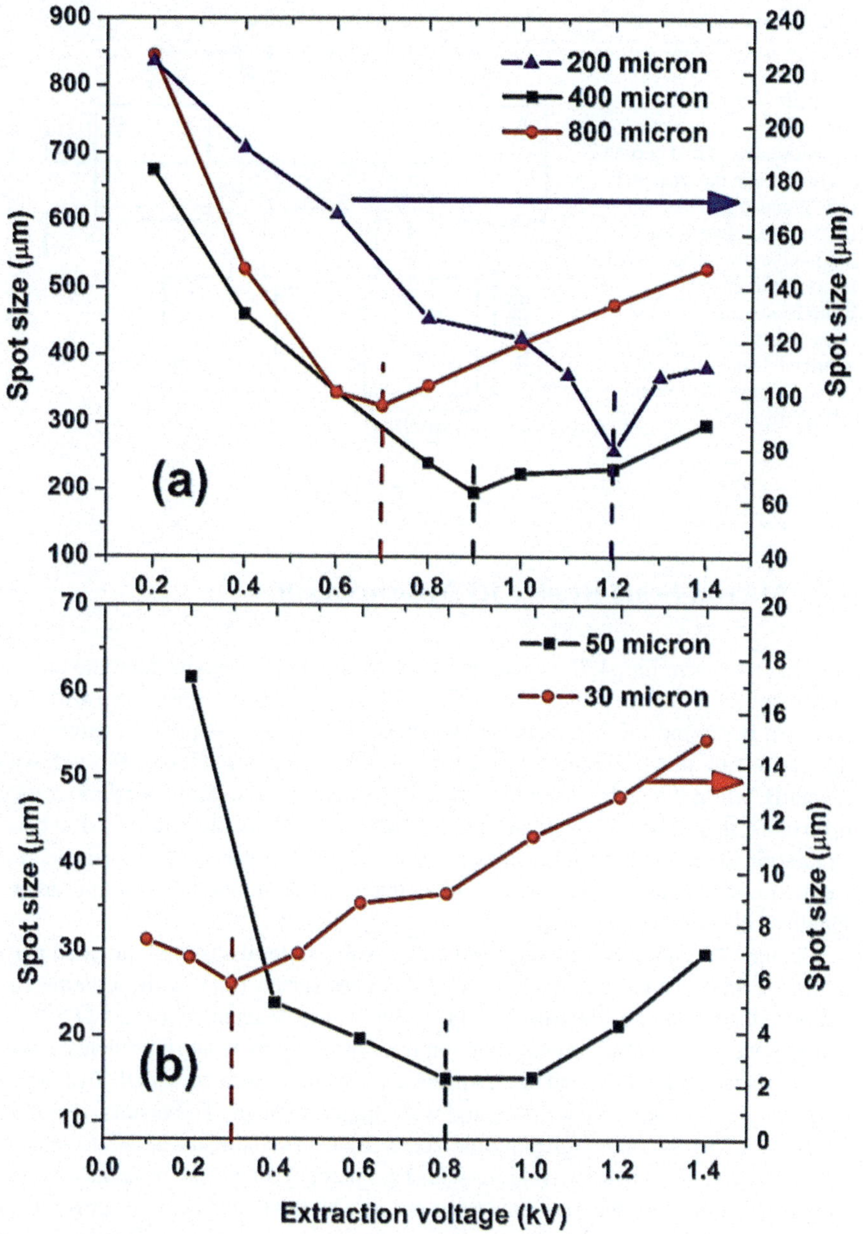

Fig. 4.6 The variation of the beam spot size with the extraction voltage **a** for bigger apertures (800, 400, and 200 μm) and **b** for smaller apertures (50 and 30 μm). Dotted lines represent corresponding focal points for each aperture diameter (AIP open access article [10] distributed under the terms of the Creative Commons CC BY license)

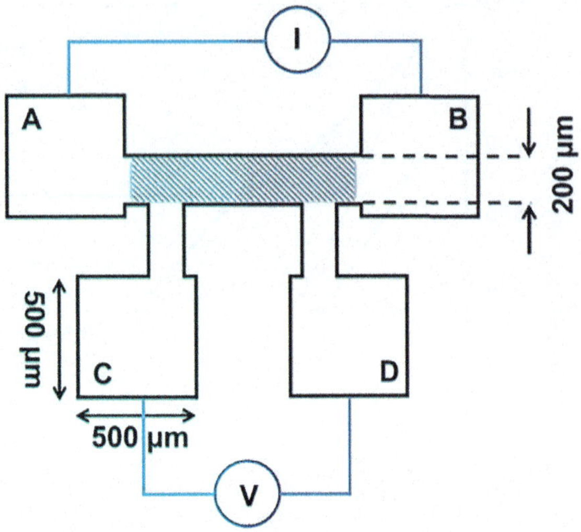

Fig. 4.7 The design of the mask for micron-scale resistance measurements. The pads A and B are connected to a constant current source so that current *I* flow through the strip and pads C and D will measure the corresponding voltage drop in the strip. The shaded region represents the beam-irradiated area (AIP open access article [10] distributed under the terms of the Creative Commons CC BY license)

4.4 Micron-Scale Resistivity Measurement

A mask, shown in Fig. 4.7, is designed for thin-film deposition to track the change in resistance at micron scale. The mask consists of a 200-μm-thick strip with four pads (A, B, C, and D) for contacts. The shape of the pads is square in nature with dimension 500 μm × 500 μm. A constant current source is connected through pads A and B, while the pads C and D were used to measure the corresponding voltage drop in the strip. The strip, with an area of 200 μm × 800 μm, is exposed to ion beam irradiation in the shaded region shown in Fig. 4.7. By comparing the resistance after irradiation with that of the pristine sample, the variation in resistance at micrometer length scales can be quantified.

Cu thin-film samples were prepared on glass substrates using the fabricated mask and a thermal evaporation system. The sheet resistance (R_S) of the pristine (as-deposited) samples was measured and found to be nearly identical (~1.25 Ω). These samples were then subjected to argon ion beam irradiation at varying fluences, with the energy fixed at 0.5 kV, and the resulting changes in R_S were recorded. To enhance sensitivity, the samples were irradiated with higher fluences. The results, including the specific fluence values, are summarized in Table 4.1. Using the MIBS technique, an array of localized resistive regions with a diameter of 30 μm can be created, where the sheet resistance of the irradiated areas is 2–6% higher than that of the unirradiated regions. The resistance of these areas can be further adjusted by varying the fluence and altering the mass of the ionic species used for irradiation (Fig. 4.8).

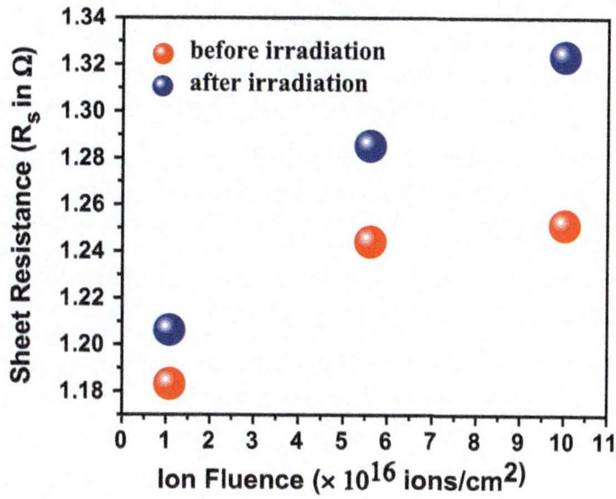

Fig. 4.8 Variation of magnitude of sheet resistance R_S with fluence (before and after irradiation) in micron-scale

Questions

1. How is the stopping potential defined for ion beams? What could be the reason for the nonlinear behavior of the total extracted beam current with the stopping potential?
2. For the three different fluences shown in Table 4.1, can you calculate the number of ions hitting a spot of diameter 37.3 microns and 77.6 microns, as shown in Fig. 4.5?
3. Find the Debye length for an Argon plasma having electron temperature $T_e =$ 10 eV and plasma density $n = 10^{11}$ cm^{-3}. Can you compare the obtained Debye length with the plasma electrode apertures mentioned in Fig. 4.6?
4. Compare and contrast the method to calculate the change in resistance in Fig. 4.7 for a Cu strip, by flowing a current and measuring the voltage drop, as compared to usual method for determination of Ohms law in a wire (for example). For a metal strip what kind of resistance is obtained?

References

1. W. Li, R. Minev, S. Dimov, G. Lalev, Patterning of amorphous and polycrystalline Ni78B14Si8 with a focused-ion-beam. Appl. Surf. Sci. **253**, 5404 (2007)
2. E. Platzgummer, H. Loeschner, G. Gross, Projection maskless patterning for nanotechnology applications. J. Vac. Sci. Technol. B **26**, 2059 (2008)
3. Y. Lee, W.A. Barletta, K.N. Leung, V.V. Ngo, P. Scott, M. Wilcox, N. Zahir, Multi-aperture extraction system with micro-beamlet switching capability. Nucl. Instrum. Methods Phys. Res. A **474**, 86 (2001)
4. J.-H. Kim, B. Tsuchiya, S. Nagata, T. Shikama, Effects of ion beam surface modification on proton conductivity of $BaCe_{0.9}Y_{0.1}O_{3-\delta}$. Solid State Ionics **179**, 1182 (2008)

5. H.-J. Maas, J. Heimel, H. Fuchs, U.C. Fischer, J.C. Weeber, A. Dereux, Photonic nanopatterns of gold nanostructures indicate the excitation of surface plasmon modes of a wavelength of 50–100 nm by scanning near-field optical microscopy. J. Microsc. **209**, 241 (2003)
6. A. Chowdhury, S. Bhattacharjee, Micron-focused ion beamlets. J. Appl. Phys. **107**, 093307 (2010)
7. A. Chowdhury, S. Bhattacharjee, Localized subsurface modification of materials using micro-low-energy multiple ion beamlets. AIP Adv. **1**, 042150 (2011)
8. S.K. Maurya, S. Barman, N. Pan, S. Bhattacharjee, Effect of plasma and beam parameters on focal dimensions in micrometer charged particle optics: enhanced nonlinear demagnification below the Debye length. Phys. Plasmas **26**, 63103 (2019)
9. S. Barman, S.K. Maurya, S. Bhattacharjee, Experimental realization of nonlinear demagnification in plasma-based charged particle optics. Plasma Res. Express **4**, 025003 (2022)
10. A. Chowdhury, S. Chatterjee, A. Dutta, S. Bhattacharjee, Stopping potential and ion beamlet control for micro-resistive patterning through sub-Debye length plasma apertures. AIP Adv. **4**, 127127 (2014)

Chapter 5
Creation, Characterization, and Applications of Atomically Heterogeneous Systems

Low-energy ion beam irradiation has become an indispensable tool across a broad spectrum of fundamental science, applied research, and materials development. Over time, ion beams spanning a wide energy range—from just a few electron volts up to several million electron volts—have been leveraged for a diverse array of research and application purposes. The specific energy of the ion beam is a critical parameter, as it determines the depth of penetration within a target material, enabling precise modification of select layers by tuning the beam energy accordingly. Furthermore, the flexibility to employ various ionic species tailored to particular applications further enhances the versatility of this technology. Ultimately, the primary factor governing the interaction between ions and a material is the energy of the incident ion beam. The behavior of low-energy ion beams differs significantly from that of high-energy ion beams. High-energy ion beams exhibit capabilities such as sputtering [1], channeling [2], deep material deposition [3], and backscattering [4, 5], driven by their deeper penetration range within materials. In contrast, low-energy ion beams primarily interact with surface and subsurface regions, making them ideal for applications like surface modification, nanopatterning, and inducing subsurface excitations [6–24]. Given their limited penetration depth, low-energy ion beams are uniquely suited to applications that require precise surface and near-surface modifications, offering unparalleled control over material properties at the nanoscale. This distinction underscores their critical role in advancing technologies that demand atomic-scale precision.

In recent years, low-energy ion beams have garnered significant attention due to their unique interactions with matter, particularly in the precise manipulation of subsurface atomic layers for a wide range of research applications [6–9, 21, 22, 25–27]. These ion beams serve as a powerful tool for non-destructive studies of surface physics. As highlighted above, low-energy ion beams have been extensively employed for surface modification [21, 25–27] and patterning, including the formation of surface ripples [6–9]. Additionally, studies have focused on hot carrier excitations in gold thin films irradiated by low-energy alkali and noble gas ions [31].

S. Bhattacharjee, *Atomically Heterogeneous Systems*,
https://doi.org/10.1007/978-981-95-9733-8_5

Low-energy ion beams can effectively modify various surface properties such as morphology [6–9], electronic characteristics [6–9, 23, 28], optical behavior [22, 29–32], and thermal properties of materials. Ion bombardment of clean single-crystal metal surfaces is particularly effective in enhancing surface reactivity by creating vacancies and interstitials, which increase the number of active sites and may improve catalytic performance [33, 34]. While ion irradiation-induced damage to solids, such as interactions between implanted atomic species and vacancies, self-interstitials, and dislocations, has been extensively studied in the context of high-energy ion irradiation (e.g., MeV-range hydrogen and helium ion irradiation for nuclear technology applications), there is also growing interest in the effects of medium-energy ion irradiation. Notably, changes in the resistance of metals due to the introduction of metallic impurities, such as dilute metal alloys [35, 36], or by implanting medium-energy (>10 keV) metal ions into pure samples have been explored in recent studies.

This chapter explores the use of low-energy ion beams to create atomically heterogeneous systems. These systems are formed by irradiating substrates with low-energy ion beams, which penetrate into the lattice of the host material, reaching subsurface layers typically ranging from a few to tens of nanometers in depth. In the majority of the experiments discussed in this book, metallic surfaces were used as substrates. Experimental findings show that ion beams with energies ranging from 0 to 5 keV successfully enter the subsurface layers of these metallic hosts, resulting in the formation of atomically heterogeneous surfaces. Before delving into the creation and applications of such systems, it is important to first define atomically heterogeneous systems and understand the methods for their experimental realization in the laboratory.

5.1 What is an Atomically Heterogeneous System?

The defining characteristic of an atomically heterogeneous system is the presence of heterogeneity at the fundamental atomic level, specifically on the surface of the material. This heterogeneity arises from the interactions between the material's surface and the incident ion beams during irradiation. Notably, in most experiments, ions of inert gases such as argon, krypton, and neon are employed. These ions, upon penetration into the lattice of the host material, do not engage in chemical bonding with the substrate atoms. As a result, the surface remains chemically homogeneous, despite the atomic-scale heterogeneity introduced through physical modifications. This distinction is critical: while the surface is atomically heterogeneous due to localized structural and compositional changes, it retains chemical homogeneity because no new compounds are formed. This unique combination of atomic heterogeneity alongside chemical homogeneity is a hallmark of surfaces prepared using low-energy inert gaseous ion irradiation. Such surfaces exhibit remarkable versatility and have been shown to be valuable in a wide range of applications. For instance, ion irradiation can be used to tailor surface wettability, enabling precise control over hydrophobic or hydrophilic properties. Similarly, electrical characteristics such as surface resistivity and field emission behavior can be fine-tuned. The optical properties, including reflectivity and light absorption, can also be modified to suit specific needs. The unique surface properties that emerge in atomically heterogeneous systems, are quite different from the bulk material. These alterations are confined to the near-surface region, where ion-induced modifications occur. In this book, we will explore several novel properties and functionalities that arise in materials upon irradiation with low-energy ion beams. The case studies and examples provided will underscore how ion beam engineering enables the development of atomically heterogeneous materials with tailored properties for advanced technological applications (Fig. 5.1).

5.2 How Can Atomically Heterogeneous System Be Realized Experimentally in the Laboratory?

The formation of an atomically heterogeneous system can be experimentally realized using a specialized setup, as schematically depicted and elaborated in detail in Fig. 3.1 of Chap. 3. The experimental apparatus employs a microwave plasma source, through which the plasma is generated within a magnetic multicusp configuration created by permanent magnets. From this plasma, low-energy ion beams are extracted using custom-designed electrode system, allowing for the creation of either single broad ion beams or micron-scale single and multiple ion beamlets. The design of the electrode plays a pivotal role in beam extraction, enabling precise control over the ion beam dimensions and focusing capabilities. For instance, micron-sized ion beamlets can be effectively focused by optimizing the geometry of the plasma electrode. To generate an atomically heterogeneous system, the extracted ion beams are directed onto a substrate positioned 1–2 mm away from the exit aperture of the multicusp plasma source. The substrates typically consist of metal foils or thin metallic films, which are irradiated with a variety of inert gas ions, including argon (Ar), krypton (Kr), and neon (Ne), as well as molecular gas ions like nitrogen (N_2) and hydrogen (H_2). Importantly, while inert gas ions primarily induce structural and compositional modifications, molecular gas ions, such as nitrogen and hydrogen, may interact chemically with the host metal surface, potentially forming new compounds. The choice of substrate material spans a range of metals, including copper (Cu), aluminum (Al), gold (Au), silver (Ag), and titanium (Ti), to explore diverse interactions and outcomes. Experimental parameters such as ion species, substrate composition, ion beam fluence, and energy are systematically varied to tailor the properties of the atomically heterogeneous systems. For example, adjusting the ion beam fluence influences the extent of atomic-scale mixing, while varying the ion energy impacts defect creation and penetration depth. This experimental methodology provides a versatile platform for achieving controlled atomic heterogeneity, flagging the way for advanced material design and applications in areas such as catalysis, surface engineering, and nanostructure fabrication.

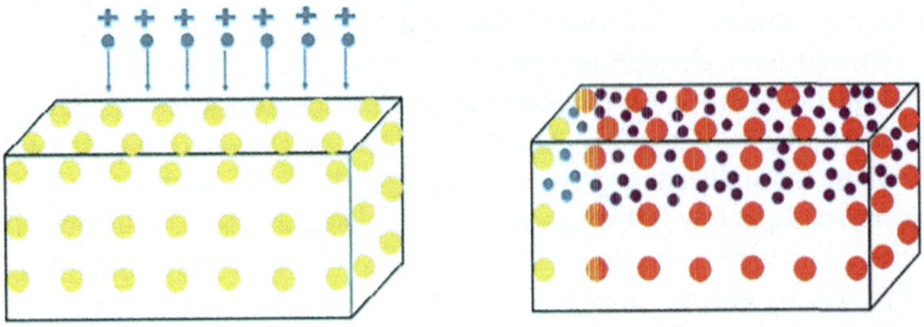

Fig. 5.1 Schematic diagram to show ion beam irradiation and the formation of atomically heterogeneous system. The red dots indicate host atoms and the purple dots indicate ions that get embedded on the substrate at the few subsurface layers

5.3 Characterization of Atomically Heterogeneous Systems

In order to better understand the physical properties and growth process of atomically heterogeneous samples, the following characterization techniques are employed.

5.3.1 Diagnostics for Material Surface and Subsurface Characterization

Cutting-edge techniques were leveraged to thoroughly examine the material surfaces generated via low-energy ion beam irradiation. The methods outlined below were strategically selected to offer a holistic understanding of the structural, chemical, and physical characteristics of the irradiated surfaces:

(1) Morphological Imaging of Surface

(a) Scanning electron microscopy (SEM), (b) scanning probe microscopy, especially atomic force microscopy (AFM), (c) transmission electron microscopy (TEM), and (d) scanning electron microscopy (SEM).

(2) Chemical Composition of the Surface and Near Surface

X-ray photoelectron spectroscopy (XPS) offers valuable insights into the composition of implanted ions and surface impurities on irradiated materials, which can significantly influence the resulting surface properties. This surface-sensitive technique provides information up to a depth of approximately 3λ, where λ represents the wavelength of the incident X-ray radiation. XPS provides information up to the depth 3λ, where λ being the wavelength of the incident X-ray.

(3) Alteration of Work Function

Kelvin probe force microscopy (KPFM) is a powerful technique used to investigate variations in contact potential difference arising from surface structural changes and ion implantation processes. This approach can provide valuable insights into the mechanisms underlying surface charging phenomena in space environments.

(4) Structural and Electronic Alteration of Surface
Additional applicable techniques like noble gas mass spectrometry, X-ray fluorescence spectroscopy (XRF), X-ray diffraction (XRD), etc. are also used to understand the growth process.

Details of the Equipment Used for Characterization

The equipment utilized throughout the investigation are described below with their detailed specifications. The model of scanning electron microscope (SEM) used was Zeiss SUPRA40VP. This model of SEM captures high-resolution micrographs of the surface topography for both pristine and ion-irradiated samples. It features a maximum resolution of 1 nm and provides a magnification range from 100× to 900,000×. The instrument can operate with an electron acceleration voltage of up to 20 kV, enabling a probing depth of a few microns from the sample surface. Consequently, SEM, along with energy-dispersive X-ray spectroscopy (EDX), provides comprehensive information spanning the entire film thickness. The transmission electron microscope (TEM) used for this work is the FEI Tecnai G2 12 Twin model, which operates at a voltage range of 20–120 kV. It employs a LaB6 emitter as the electron source and offers a line resolution of 0.2 nm with a maximum eucentric tilt angle of ±70°. TEM sample preparation involved mechanically grinding the substrate into a circular disk with a diameter of 3 mm and a thickness of approximately 100 μm. The disk was then subjected to dimpling to achieve uniform thinning. To further reduce the thickness to less than 40 μm at the center, ion milling was performed. This ultra-thin central region was used for detailed TEM analysis.

Atomic force microscopy (AFM) measurements were performed using an Asylum Research MFP-3D Classic AFM system equipped with a model ASYELEC 01 cantilever. The cantilever, made of n-type silicon coated with Ti-Ir layer, was used to scan a 25 μm^2 area of the substrate surface at different position, providing high-resolution topographical data. Additionally, Kelvin probe force microscopy (KPFM) was conducted using the same AFM setup to map variations of contact potential difference to estimate the work function of top surface and subsurface.

X-ray photoelectron spectroscopy (XPS) was carried out using a Versa Probe II spectrometer (Model-5000, FEI, Inc.). The X-ray beam provided a spatial resolution of approximately 200 μm, and the sampled area was around 1 mm^2. The Al Kα radiation (of energy 1486.6 eV) was used as an X-ray source and the energy resolution of the detector was $\Delta E \leq 0.5$ eV. During the experiment the chamber was kept at ultra-high vacuum conditions (~10^{-9} mbar). Calibration of XPS spectra was performed using the carbon C1s core-level binding energy reference at 284.8 eV. Peak fitting was conducted without imposing any constraints, ensuring accurate resolution beyond the detector's limits. Elemental concentrations, including implanted ions, were quantified to the first decimal place using high-resolution spectra after normalization by sweep and sensitivity factor with the help of Casa-XPS software [37]. During the XPS measurement no depth profiling was carried out. These advanced characterization methods provided detailed structural, compositional, and electronic information, ensuring accurate analysis of the synthesized materials [38].

5.3.2 *Optical Characterization*

(a) *Spectrophotometer for Measurement of Reflectivity, Transmissivity, and Absorptivity*

The atomically heterogeneous systems possess different optical properties compared to the bulk material. The reflection, transmission, and absorption of atomically heterogeneous systems were measured using a Perkin Elmer Lambda 950 UV-VIS-NIR spectrophotometer. This instrument covers a broad wavelength range from 175 nm to 3300 nm, providing exceptional versatility for analyzing optical behavior across ultraviolet (UV), visible (VIS), and near-infrared (NIR) regions. The spectrophotometer offers high-resolution capabilities, ranging from 0.05 nm to 5.00 nm in the UV/VIS spectrum and 0.20 nm to 20.00 nm in the NIR region. Reflectivity and transmissivity measurements were conducted for both pristine and ion-irradiated samples at an angle of 8° relative to surface normal. This angle was chosen to optimize measurement accuracy while minimizing errors due to surface scattering of UV-VIS-IR photon. The design of the spectrophotometer ensures precise transmissivity measurements, with an instrumental uncertainty of approximately $<7 \times 10^{-5}\%$ over the investigated wavelength range. To maintain sample integrity and prevent contamination, all measurements were performed under ambient atmospheric conditions. The samples were carefully stored in a vacuum desiccator (roughly 10^{-2} torr) and retrieved only immediately before optical characterization. For transmissivity experiments involving thin metallic films, the films were deposited onto glass substrates. This ensured compatibility with the optical system and minimized interference effects that might arise from substrate-related mismatches. Absorptivity (A) was calculated using the principle of energy conservation, expressed as $R+T+A = 1$, where R and T denote the experimentally measured magnitude of reflectivity and transmittance, respectively. This approach provided a comprehensive understanding of the energy interactions at the surface of the materials, enabling precise determination of optical behavior. The meticulous methodology adopted in these measurements ensured accurate characterization of the optical properties of the synthesized atomically heterogeneous systems. These findings contribute valuable insights into the tunable optical responses of materials subjected to ion irradiation, showcasing their potential for various advanced applications.

(b) *Pseudo-Brewster Angle Technique for Measurement of Refractive Index*

The real (n) and imaginary (k) components of the refractive index were estimated using the pseudo-Brewster angle technique, as detailed in Ref. [39]. To achieve this, an advanced experimental setup has been meticulously designed and is schematically illustrated in Fig. 5.2. This methodology provides a precise means of analyzing the optical properties of materials, particularly pristine and ion-irradiated metallic thin films (MTFs). In this setup, the sample is mounted vertically on a highly precise angular rotary stage with a resolution of 0.5°. This stage allows for controlled and accurate adjustment of the sample's orientation during the experiment. The incident light source is a laser capable of generating either unpolarized or semi-polarized light at a specific wavelength, depending on the experimental requirements. The laser beam first passes through a linear sheet polarizer, which ensures that the transmitted light is horizontally polarized along the optical axis. The horizontally polarized light then enters a half-wave plate (HWP). The HWP is carefully adjusted to rotate the polarization of the transmitted beam by 45°, thereby splitting the light into two orthogonal components: horizontally polarized (p-polarized) and vertically polarized (s-polarized) light, with equal intensity. The p-polarized light component is subsequently reflected from the surface of the sample, which is positioned on the angular rotary stage. After reflection, the light passes through an optical analyzer, strategically placed between the sample and the detector. The analyzer is employed to selectively analyze the polarization state of the reflected light. Both the detector and the analyzer are mounted on a movable arm, which is part of a spectrometer system. During the experimental procedure, the laser, polarizer, and HWP remain stationary, ensuring a consistent light source and polarization configuration throughout the measurements. The movable arm, which houses the detector and analyzer, is rotated in conjunction with the sample's angular adjustments. This synchronized movement enables the measurement of reflectivity at various angles of incidence. The reflectivity of the p-polarized light (R_P) is recorded as the detector is positioned at incremental angles. At the pseudo-Brewster angle the reflectivity of the p-polarized light reaches a minimum. This angle is a critical parameter used to extract the optical constants of the sample. Reflectivity measurements are systematically performed for both pristine and ion-irradiated metallic thin films, allowing for a comparative analysis of the effects of ion irradiation on the material's optical properties. The optical constants (n) and (k) are subsequently calculated using established mathematical relationships, which are discussed in the following sections. This meticulous experimental setup ensures high accuracy and reproducibility in determining the refractive index components. It provides valuable insights into the optical behavior of advanced materials, particularly those subjected to ion beam modifications [40]. To find out the optical constants, the following equations are used:

$$n = \frac{1 - R_{PB}}{1 + R_{PB}} \sec\theta_B, \tag{5.1}$$

and

$$k = \frac{2\sqrt{R_{PB}}}{1 + R_{PB}} \sec\theta_B. \tag{5.2}$$

where θ_B is the pseudo-Brewsters angle and R_{PB} is the corresponding reflectivity of p-polarized light, which are measured from the reflectivity versus incidence angle plot, and are then employed to obtain the optical constants of MTF using the above Fresnel's equations for the air-metal interface.

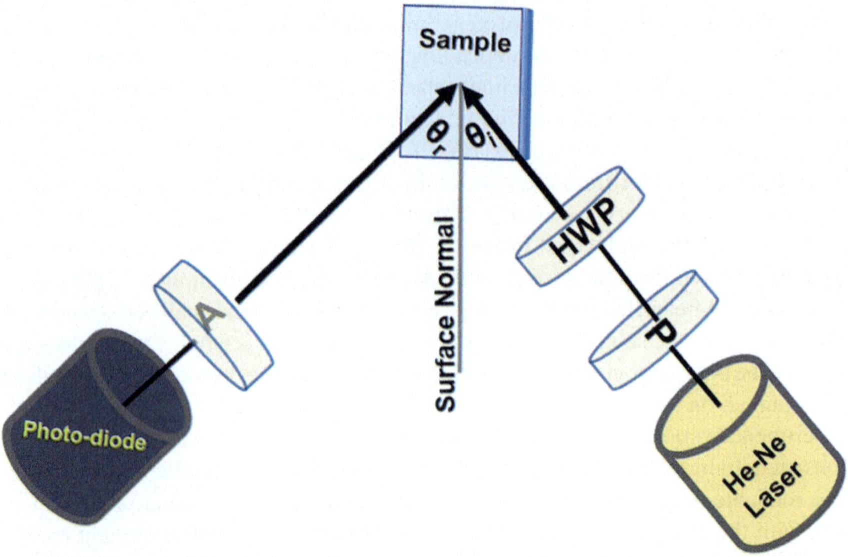

Fig. 5.2 Schematic diagram of optical setup for measurement of optical constants. P, linear sheet polarizer; HWP, half-wave plate; A, linear sheet analyzer; He–Ne laser; Si photodiode (reprinted with permission from [39] © Optica Publishing Group)

5.3.3 Electrical Characterization

(a) *Van der Pauw Technique for Measurement of Sheet Resistance*

The atomically heterogeneous system leads to an altered magnitude of sheet resistance. In order to measure the sheet resistance of the films, van der Pauw technique [41–43] was used. The schematic diagram presented in Fig. 5.3 exhibits how the current and voltages are measured across different directions. The measured resistances, R_1 and R_2, along two perpendicular directions (as shown in Fig. 5.3) can be used to measure the sheet resistance (R_S) using the formula:

$$e^{-\pi R_1/R_S} + e^{-\pi R_2/R_S} = 1, \tag{5.3}$$

where the resistivity ρ is given by $\rho = R_S \times d$, d being the thickness of the thin film. Further, $|R_1 - R_2|$, which is a measure of homogeneity of the film, was found to be ~ 80%.

(b) *Field Emission Measurement Setup*

The emission of electrons from the atomically heterogeneous surface, under the application of electric field, known as field emission spectroscopy, was investigated using a specialized, custom-built experimental setup designed for the operation at room temperature. This setup was created from the terminal section of a plasma-based

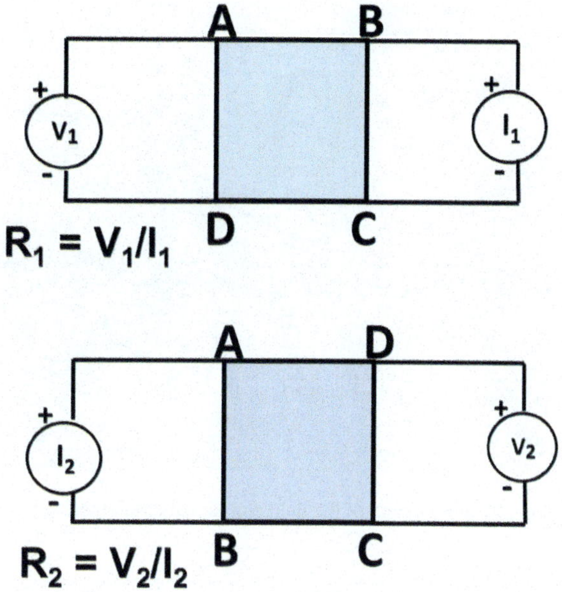

Fig. 5.3 Schematic diagram of van der Pauw method. The square film is demarcated by ABCD and the current is applied and voltage is measured in the two directions as shown (reprinted from [Abhishek Chowdhury and Sudeep Bhattacharjee, AIP Advances **1**, 042150 (2011)], with the permission of AIP)

focused ion beam system that was developed in-house, as described in previous experiments [44, 45]. A sample holder equipped with an XYZ manipulator formed a critical part of the system, enabling precise positioning and alignment of the samples (Fig. 5.4). The layout of the system, including the manipulator, is depicted in the digital image shown in Fig. 5.5. The experimental apparatus featured a parallel plate configuration with a fixed inter-electrode gap of 2 mm. This arrangement was housed within a high-vacuum chamber, which was maintained at a base pressure of approximately 1.33×10^{-6} mbar to prevent contamination and ensure accurate measurements. The detailed schematic of this configuration is illustrated in Fig. 5.5a, while a corresponding digital photograph of the actual setup is displayed in Fig. 5.5b. During the measurements, the collector holder was specifically designed to cover only the uniformly irradiated central circular region of the substrates. This design effectively eliminated any edge-related emissions, thereby ensuring that the measurements were solely reflective of the central emission properties of the synthesized structures. The anode used in the setup was a mirror-polished copper plate, while the cathode was the substrate itself, providing a clean and stable interface for electron field emission. To maintain electrical isolation and prevent interference, both the collector and sample holders were fabricated from Teflon, a non-conducting material. This careful design choice minimized parasitic currents and ensured that the field emission measurements were both precise and reliable. The experimental setup thus offered a robust and effective means to characterize the electron emission properties of the synthesized atomically heterogeneous materials under controlled conditions.

Fig. 5.4 Photograph of the field emission measurement setup

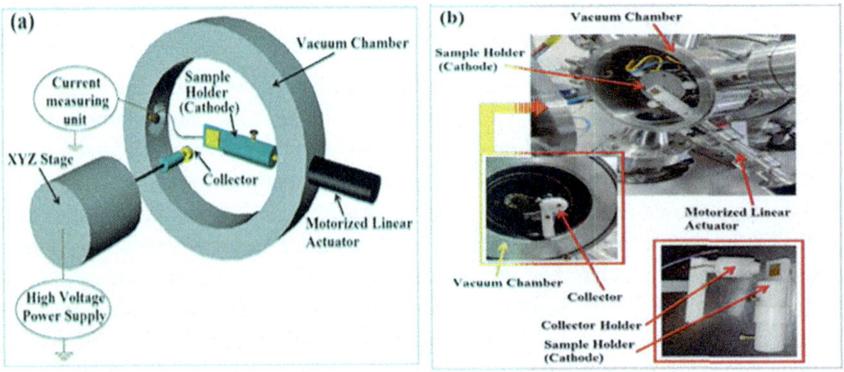

Fig. 5.5 **a** Schematic diagram of an experimental setup for a field emission measurement experiment and **b** a digital image of the field emission system (not to scale) (reprinted from [Jayashree Majumdar and Sudeep Bhattacharjee, J. Appl. Phys. 132, 083304 (2022)], with the permission of AIP Publishing)

5.4 Experiments Carried Out with Atomically Heterogeneous Systems

The low-energy ion beams are instrumental in creating atomically heterogeneous systems. These systems can be formed on a wide range of material surfaces, including conductors, semiconductors, insulators, and even biological substrates. While the projectile ions can originate from any gas species, the majority of the studies presented in this work utilize inert gaseous ions as projectiles. Following their creation, these systems have been extensively studied to investigate various physical properties of ion beam-modified materials, including

1. Electrical properties: Modifications in sheet resistivity of metallic thin films, evaluated at both room and low temperatures, provide insights into temperature-dependent resistivity in metals. These studies also explore the influence of ionic impurities on parameters such as the Debye temperature and elastic modulus.
2. Surface morphology and work function: The introduction of ionic impurities leads to the formation of diverse surface structures. These structural modifications have been linked to significant changes in work function of the material.
3. Optical properties: Ion beam irradiation induces changes in the optical characteristics of materials, including alterations in the real and imaginary parts of the refractive index (dielectric constants), skin depth, and optical conductivity.
4. Field emission properties: Surface micro-conical structures created through ion beam modification are utilized to investigate enhanced field emission behavior, revealing the impact of ion irradiation on emission efficiency and stability.

All of the above-mentioned topics are explored in detail in the following chapters, providing comprehensive insights into the transformative effects of ion beam modification on material properties.

Questions

1. Distinguish between the effect of high-energy and low-energy ion beam interaction with matter.
2. What are the different types of effects low-energy ion beams can bring about on the materials with which it interacts? Mention the effects with references.
3. What is an atomically heterogeneous system? What are the advantages of using inert gaseous ions?
4. What are the experimental parameters that can be varied in the experiment on the investigation of creation of atomically heterogeneous systems with ion beams?
5. What are the three main diagnostic techniques to characterize the surface morphology of the ion beam-irradiated materials? Compare and contrast these different techniques.
6. What information from the subsurface layers of the material can be obtained by employing X-ray photoelectron microscopy (XPS)?
7. Which instrument can be utilized to measure the optical properties such as reflectivity, transmissivity, and absorptivity of metallic thin films? Why would these properties depend upon the wavelength of incident light?
8. What are the functions of a polarizer and a half-wave plate in the pseudo-Brewster angle technique for the measurement of the real and imaginary parts of the refractive index?

9. What is meant by Brewster angle? Why is the technique mentioned in this chapter for the measurement of the real and imaginary parts of the refractive index called pseudo-Brewster angle technique? Why does the reflectivity of the p-polarized light reach a minimum at the pseudo-Brewster angle?
10. What is sheet resistance of a metallic thin film? What is the unit of sheet resistance? What method is applied to measure the sheet resistance?
11. What is meant by field emission of a material? What is the governing theory behind the prediction of the field emission current of a material?
12. A thin film of indium tin oxide (ITO) is deposited on a glass substrate, and its reflectivity is measured at various angles of incidence for p-polarized light of wavelength $\lambda = 632.8$ nm (He–Ne laser). The pseudo-Brewster angle (θ_p) is observed at which the reflectivity reaches a minimum but does not go to zero due to the complex refractive index of ITO. If the complex refractive index of ITO: $n_{ITO} = 1.8 + i0.02$, refractive index of glass substrate: $n_s = 1.52$ and $\theta_p \approx \tan^{-1}\{\mathrm{Re}(n_{ITO}/n_s)\}$, (a) Calculate the pseudo-Brewster angle (θ_p) for the ITO/glass interface. (b) Discuss why the reflectivity does not drop to zero at θ_p and how the absorptive nature of ITO influences the reflectivity minimum. (c) If the extinction coefficient (k) of ITO increases, how would the pseudo-Brewster angle and minimum reflectivity change?
13. How does the pseudo-Brewster angle technique help in determining the refractive index of a thin film, and why does the reflectivity for p-polarized light reach a minimum at this angle instead of zero?

References

1. R.T. Yadav, V.S. Ramamurthy, D.P. Mahapatra, G.V. Raviprasad, Size distribution of atomic clusters formed by energetic-heavy-ion sputtering. Phys. Rev. A **49**, 1908 (1994)
2. W.R. Wampler, Transmission-ion-channeling studies of the silicon (111) monohydride surface. Phys. Rev. B **55**, 9693 (1997)
3. T. Roller, W. Bolse, Oxygen diffusion and oxide phase formation in iron under swift heavy ion irradiation. Phys. Rev. B **75**, 054107 (2007)
4. L.C. Northcliffe, Energy loss and effective charge of heavy ions in aluminum. Phys. Rev. **120**, 1744 (1960)
5. W. Schimmerling, K.G. Vosburghand, P.W. Todd, Measurements of range in matter for relativistic heavy ions. Phys. Rev. B **7**, 2895 (1973)
6. S. Macko, F. Frost, B. Ziberi, D. Forster, T. Michely, Is keV ion-induced pattern formation on Si(001) caused by metal impurities? Nanotechnology **21**, 085301 (2010)
7. B. Ziberi, F. Frost, T. Hoche, B. Rauschenbach, Ripple pattern formation on silicon surfaces by low-energy ion-beam erosion: experiment and theory. Phys. Rev. B **72**, 235310 (2005)
8. P.F.A. Alkemade, Propulsion of ripples on glass by ion bombardment. Phys. Rev. Lett. **96**, 107602 (2006)
9. H. Hansen, A. Redinger, S. Messlinger, G. Stoian, Y. Rosandi, H.M. Urbassek, U. Linke, T. Michely, Mechanisms of pattern formation in grazing-incidence ion bombardment of Pt(111). Phys. Rev. B **73**, 235414 (2006)
10. I. Koponen, M. Hautala, O.-P. Sievanen, Simulations of ripple formation on ion-bombarded solid surfaces. Phys. Rev. Lett. **78**, 2612 (1997)

11. Q. Wei, J. Lian, L.A. Boatner, L.M. Wang, R.C. Ewing, Propagation of ripples on pyrochlore induced by ion beam bombardment. Phys. Rev. B **80**, 085413 (2009)
12. R. Moroni, D. Sekiba, F. Buatier de Mongeot, G. Gonella, C. Boragno, L. Mattera, U. Valbusa, Uniaxial magnetic anisotropy in nanostructured co/cu (001): From surface ripples to nanowires. Phys. Rev. Lett. **91**, 167207 (2003)
13. P. Chaudhari et al., Atomic-beam alignment of inorganic materials for liquid-crystal displays. Nature **411**, 56 (2001)
14. L. Vattuone, U. Burghaus, L. Savio, M. Rocca, G. Costantini, F. Buatier de Mongeot, C. Boragno, S. Rusponi, U. Valbusa, Oxygen interaction with disordered and nanostructured ag(001) surfaces. J. Chem. Phys. **115**, 3346 (2001)
15. S. Facsko, T. Dekorsy, C. Koerdt, C. Trappe, H. Kurz, A. Vogt, H.L. Hartnagel, Formation of ordered nanoscale semiconductor dots by ion sputtering. Science **285**, 1551 (1999)
16. R. S. Dhaka, C. Biswas, A. K. Shukla, S. R. Barman, and Aparna Chakrabarti, Xe and Ar nanobubbles in Al studied by photoemission spectroscopy, Phys. Rev. B 77, 104119 (2008)
17. P.B. Johnson, D.J. Mazey, Gas-bubble superlattice formation in bcc metals. J. Nucl. Mater. **218**, 273 (1995)
18. R.S. Dhaka, S.R. Barman, Plasmon excitations by photoelectron emission from rare gas nanobubbles in aluminum. Phys. Rev. Lett. **104**, 036803 (2010)
19. C.J. Rossouw, S.E. Donnelly, Superheating of small solid-argon bubbles in aluminum. Phys. Rev. Lett. **55**, 2960 (1985)
20. M. Schmid, W. Hebenstreit, P. Varga, S. Crampin, Quantum wells and electron interference phenomena in Al due to subsurface noble gas bubbles. Phys. Rev. Lett. **76**, 2298 (1996)
21. S. Vauth, S.G. Mayr, Ion bombardment induced smoothing of amorphous metallic surfaces: experiments versus computer simulations. Phys. Rev. B **77**, 155406 (2008)
22. D.S. Martin, R.J. Cole, P. Weightman, Effects of ion bombardment on the optical and electronic properties of Cu(110). Phys. Rev. B **72**, 035408 (2005)
23. M.P. Ray, R.E. Lake, C.E. Sosolik, L.B. Thomsen, G. Nielsen, I. Chorkendorff, O. Hansen, Subsurface excitations in a metal. Phys. Rev. B **80**, 161405. (R) (2009)
24. V.F. Reutov, A.S. Sokhatskii, Helium ion bombardment induced amorphization of silicon crystals. Tech. Phys. Lett. **28**, 615 (2002)
25. F. Frost, A. Schindler, F. Bigl, Roughness evolution of ion sputtered rotating inp surfaces: pattern formation and scaling laws. Phys. Rev. Lett. **85**, 4116 (2000)
26. J.E. Gayone, R.G. Pregliasco, G.R. Gomez, E.A. Sanchez, O. Grizzi, Topographic and crystallographic characterization of a grazing-ion-bombarded GaAs(110) surface by time-of-flight ion-scattering spectrometry. Phys. Rev. B **56**, 4186 (1997)
27. C. Steiner, B. Bolliger, M. Erbudak, M. Hong, A.R. Kortan, J. Kwo, J.P. Mannaerts, Structural modifications of the Gd_2O_3(110) films on GaAs(100). Phys. Rev. B **62**, R10614 (2000)
28. S. Khakshouri, D.M. Duffy, Influence of electronic effects on the surface erosion of tungsten. Phys. Rev. B **80**, 035415 (2009)
29. P. Schmuki, L.E. Erickson, D.J. Lockwood, Light emitting micropatterns of porous Si created at surface defects. Phys. Rev. Lett. **80**, 4060 (1998)
30. B. Panigrahy, M. Aslam, D. Bahadur, Controlled optical and magnetic properties of ZnO nanorods by Ar ion irradiation. Appl. Phys. Lett. **98**, 183109 (2011)
31. F. Everts, H. Wormeester, B. Poelsema, Optical anisotropy induced by ion bombardment of Ag(001). Phys. Rev. B **78**, 155419 (2008)
32. X.M. Goh, N.M. Dragomir, D.N. Jamieson, A. Roberts, D.X. Belton, Optical tomographic reconstruction of ion beam induced refractive index changes in silica. Appl. Phys. Lett. **91**, 181102 (2007)
33. A.S. Soltan, R. Vassen, P. Jung, Migration and immobilization of hydrogen and helium in gold and tungsten at low temperatures. J. Appl. Phys. **70**, 793 (1991)
34. A.S. Soltan, P. Jung, A.A. Gadallah, Resistivity and implantation distribution of hydrogen and helium in metals. J. Phys. Condens. Matter **4**, 9573 (1992)
35. D.L. Mills, A. Fert, I.A. Campbell, Temperature dependence of the electrical resistivity of dilute ferromagnetic alloys. Phys. Rev. B **4**, 196 (1971)

36. J.B. Dunlop, G. Gruner, F. Napoli, Phonon resistivities of dilute transition metal compounds. Solid State Commun. **15**, 13 (1974)
37. J. Mukherjee, D. Bhowmik, G. Bhattacharyya, B. Satpati, P. Karmakar, Spatially varying chemical phase formation on silicon nano ripple by low energy mixed ions bombardment. J. Phys. Condens. Matter **34**, 135001 (2022)
38. J. Walton, A. Carrick, *The Casa Cookbook: The Casa XPS User's Manual, Part 1: Recipes for XPS Data Proceedings* (Acolyte Science, 2009)
39. K.P. Singh, J. Majumdar, S. Bhattacharjee, Tuning optical properties of atomically heterogeneous systems created by plasma-based low-energy ion beams. Appl. Opt. **59**, 4507 (2020)
40. R. Sengupta, A. Adhiya, K.S.R. Sekhar, R. Kaur, Measurement of complex dielectric constant using optical method. IEEE Trans. Instrum. Meas. **68**, 1814–1820 (2018)
41. L.J. van der Pauw, Philips Res. Rep. **13**, 1 (1958)
42. G.T. Kim, J.G. Park, Y.W. Park, Y.W. Park, C. Muller-Schwanneke, M. Wagenhals, S. Roth, Nonswitching van der Pauw technique using two different modulating frequencies. Rev. Sci. Instrum. **70**, 2177 (1999)
43. A. Chowdhury, S. Bhattacharjee, Localized subsurface modification of materials using micro-low-energy multiple ion beamlets. AIP Adv. **1**, 042150 (2011)
44. S.K. Maurya, S. Bhattacharjee, Variable gaseous ion beams from plasmas driven by electromagnetic waves for nano-micro structuring: a tutorial and an overview of recent works and future prospects. Plasma Res. Express **2**, 033001 (2020)
45. S.K. Maurya, S. Barman, N. Pan, S. Bhattacharjee, Effect of plasma and beam parameters on focal dimensions in micrometer charged particle optics: enhanced nonlinear demagnification below the Debye length. Phys. Plasmas **26**, 063103 (2019)

Chapter 6
Basic Experiments in Plasma and Ion Beams with Sample Preparation Techniques

Before conducting experiments with ion beams to fabricate atomically heterogeneous systems for various physical applications, preliminary investigations were necessary to characterize the plasma and ion beams produced within the multicusp confinement device. These initial studies were crucial for understanding the plasma's properties, including density profiles and electron temperature, in order to estimate the maximum achievable ion densities and currents from the plasma source. Standard electrostatic diagnostic techniques, such as Langmuir probes, were employed to perform this plasma characterization. Subsequently, the extracted ion beam flux was measured as a function of applied microwave power and neutral gas pressure. From these measurements, the beam fluence delivered to samples could be calculated. Additionally, the uniformity of the extracted ion beam was analyzed, and polar plots of the ion saturation current at the edge of the multicusp were generated. Beyond the plasma and ion beam characterization, this chapter also details the sample preparation techniques used to create the desired atomically heterogeneous systems. These preparatory and characterization studies provided a foundational understanding to optimize the performance and applications of the atomically tailored systems.

6.1 Plasma Characteristics Inside the Multicusp: Octupole, Decapole, and Duo-Decapole

The multicusp is a compact and unique specialized magnetic arrangement of magnets that confines charged particles via the magnetic mirror effect within the plasma [1]. The multicusp creates the static magnetic field required to meet the electron cyclotron resonance (ECR) condition, enabling efficient plasma generation. To support uniform and dense plasma formation, a region of minimum magnetic field strength is strategically positioned at the center of the multicusp [2–4]. The side and cross-sectional views of three distinct multicusp configurations such as octupole (8 poles), decapole

S. Bhattacharjee, *Atomically Heterogeneous Systems*,
https://doi.org/10.1007/978-981-95-9733-8_6

(10 poles), and duo-decapole (12 poles) are illustrated below for comparative analysis. These configurations demonstrate variations in magnetic field arrangement and plasma confinement efficiency.

The multicusp device employs an arrangement of magnetic field lines, which form cusp-like regions near the magnets, leading to an increased density of magnetic field lines in their immediate vicinity. This strategic configuration is instrumental in enhancing plasma confinement efficiency. The multicusp structure is meticulously designed using Neodymium-Iron-Boron (NdFeB) permanent magnets, renowned for their high magnetic strength and durability. These magnets are encased within square aluminum pipes and arranged in a precise circular configuration to achieve a null magnetic field at the center of the device. The square cross section of the magnets ensures seamless alignment within the aluminum pipes, resulting in an optimized magnetic field geometry that enhances plasma containment [2]. The magnets generate a surface magnetic field strength of approximately 0.5 Tesla, which is critical for robust plasma confinement. This field strength not only ensures effective containment of charged particles but also facilitates the uniform formation of dense plasma within the device. The use of NdFeB magnets, combined with the structural integrity provided by the aluminum housing, underscores the device's capability to maintain stable and high-intensity magnetic fields over extended operational periods. A thorough fabrication methodology for the multicusp device, including the arrangement and installation of the magnets and aluminum housing, is comprehensively outlined in Chap. 2. This ensures reproducibility and provides insights into the design considerations that optimize the magnetic field structure for enhanced plasma performance. The present chapter focuses on analyzing the plasma properties within three distinct multicusp configurations: (a) octupole (8 poles), (b) decapole (10 poles), and (c) duo-decapole (12 poles). All the multicusps have identical dimensions, measuring 32 cm in length. Figure 6.1a–c illustrates the fabricated octupole, decapole, and duo-decapole multicusp configurations, respectively. A fabricated multicusp is shown in Fig. 6.1d. Radial magnetic field profiles for these configurations were measured using a Hall probe connected to a digital Gaussmeter (Lake Shore 421), with results presented in Fig. 6.1e. The electron cyclotron resonance (ECR) condition, essential for efficient plasma generation, is governed by the equation $\omega = \omega_c = eB/m$, where ω is the angular wave frequency, ω_c is the angular electron cyclotron frequency, e is the electronic charge, B is the static magnetic field, and m is the mass of an electron. For the 2.45 GHz microwaves used in this study, the resonance condition is achieved at a magnetic field strength of 875 Gauss. The corresponding ECR locations are approximately 2.5 cm, 3.8 cm, and 4.4 cm from the center for the octupole, decapole, and duo-decapole configurations, respectively, as shown in Fig. 6.1e. The magnetic field intensity progressively increases closer to the boundary wall of the multicusp, as evident in the plot. Plasma density measurements indicate that the octupole configuration exhibits higher plasma density than the decapole and duo-decapole configurations, demonstrating superior plasma confinement efficiency. Consequently, the octupole configuration was selected for subsequent experiments. The observed magnetic field profile closely aligns with the theoretical expression $B = B_0(r/a)^{(n/2-1)}$, where B_0 is the surface magnetic field of the magnets, a is the radius of the multicusp, n is the number of poles, and r is the radial distance from the center [5]. Interestingly, the ECR location shifts closer to the periphery as the number of poles increases in the multicusp configuration. Therefore, during experiments, the microwave power density must be maintained constant across all configurations, with adjustments made to ensure uniform power delivery. This approach ensures consistency in plasma generation and facilitates a meaningful comparison of the performance of different multicusp designs.

The density (N_e) and temperature (T_e) of electrons in different multicusp arrangements, i.e., octupole, decapole, and duo-decapole, are measured using the Langmuir probe technique, based on the current–voltage (I–V) characteristics obtained from the probe. The plasma generated at central region of the multicusp was investigated by different parameters of plasma as a function of pressure of the gas as well as microwave power. Moreover, the variation of these plasma parameters with radial distance was also investigated as a function of microwave power and gas pressure. Since the ion flux on target sample is primarily governed by the plasma density, it is therefore crucial to investigate the behavior of plasma to optimize the ion beam extraction process.

Figure 6.2a and b represents the variation of electron density (N_e) and electron temperature (T_e) obtained from radial profile of octupole multicusp configuration, at three different pressures of the gas, i.e., 0.2, 0.5, and 0.8 mTorr. The electron density

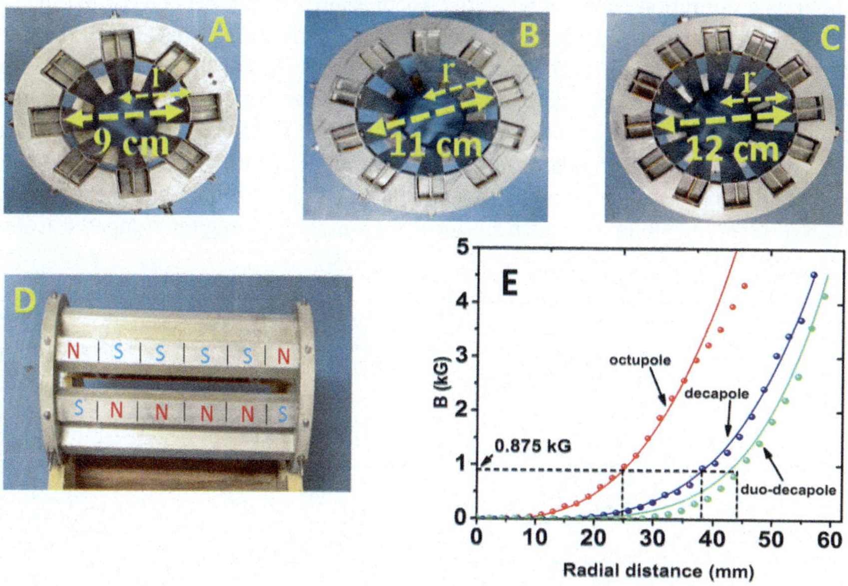

Fig. 6.1 Cross-sectional view of **a** octupole, **b** decapole, **c** duo-decapole, **d** side view with polarity arrangements of magnets inside the octupole, and **e** magnetic field variation inside octupole, decapole, and duo-decapole with radial distance from center to pole (reproduced with permission from author [Krishn Pal Singh, Optical properties of atomically heterogeneous systems created by microwave plasma-based low energy ion beams, Ph.D. thesis, IIT Kanpur, India, 2023])

(N_e) exhibits a nearly uniform distribution in the central region of the plasma, with a gradual decrease as the radial distance from the multicusp axis increases. The radial distance (r) is measured from the center, where $r = 0$, to the outer boundary of the multicusp, located at $r = 4.5$ cm for the octupole configuration. Furthermore, a decrease in electron density with increasing neutral gas pressure is observed, consistent with previous experiments. Elevated gas pressures result in more frequent collisions between electrons and neutral gas atoms, leading to energy loss by the electrons and a subsequent reduction in their ionization efficiency. The electron density in plasma is measured to be of the order of ~ 10^{16} per m^{-3}, which is adequate for the specific requirements of the experiments.

On the other hand, for octupole configuration of multicusp, plasma temperature (T_e) decreases with increase in radial distance, as observed from Fig. 6.2b. The plasma temperature (T_e) is maximum at radial distance 2.3 cm away from the center of multicusp, being closer to the position of ECR, i.e., ~ 2.5 cm from the center. It is also observed that T_e decreases with an increase in neutral gas pressure as presented in Fig. 6.2b with higher in magnitude at lower in gas pressure. Since, at a lower pressure, the number of collisions suffered by an electron is less (the mean free path of the electrons is large), this results in a larger gain in kinetic energy of the electrons before two successive collisions. The nature of variation of the electron temperature is almost similar for all the three pressures. A drastic change in temperature with change in pressure from 0.2 to 0.5 mTorr is observed from Fig. 6.2b, which also suggests a saturation behavior when the gas pressure reaches beyond 0.5 mTorr.

As depicted in Fig. 6.3, the plasma density (N_e) distribution in the decapole configuration is observed to be uniform within the central region of the plasma, mirroring the behavior seen in the octupole configuration. However, as the radial distance from the axis of the decapole increases, the plasma density gradually decreases. This decline is similar to that observed in the octupole configuration but with some notable differences. The decapole system features a larger minimum magnetic field region, resulting in an expanded uniform plasma density region compared to the

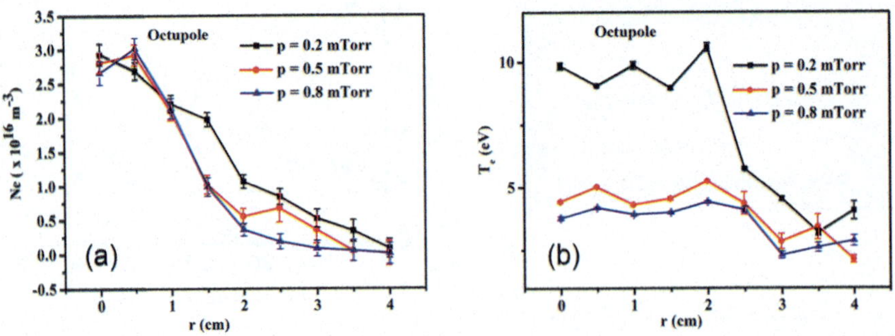

Fig. 6.2 Radial profile of **a** electron density, **b** electron temperature at three different pressures inside the octupole at microwave power of 180 W (reproduced with permission from author [Krishn Pal Singh, Optical properties of atomically heterogeneous systems created by microwave plasma-based low energy ion beams, Ph.D. thesis, IIT Kanpur, India, 2023])

octupole. This characteristic makes the decapole configuration advantageous for applications requiring uniform ion extraction, as ions are primarily extracted from the central plasma region due to its uniformity. As in the case of the octupole, the plasma density decreases with an increase in neutral gas pressure. This reduction can be attributed to the increased frequency of collisions between electrons and neutral gas atoms, leading to energy loss in the electrons and a diminished ionization rate. The overall magnitude of the plasma density in the decapole configuration is comparable to that of the octupole configuration. However, the rate of plasma density decrease is more pronounced between $r = 1.5$ cm and $r = 4$ cm, after which the density stabilizes and becomes nearly constant. In addition to plasma density, the electron temperature T_e exhibits a notable radial variation. The electron temperature decreases with increasing radial distance, reaching a peak value at approximately $r = 3$ cm from the multicusp center. This peak coincides with the position of electron cyclotron resonance (ECR), located around (3.7 cm) from the center. The trend of variation of T_e as a function of neutral gas pressure is consistent with that observed in the octupole configuration, as illustrated in Fig. 6.3b. Lower gas pressures result in higher electron temperatures due to reduced collisional damping, which allows electrons to retain more energy from the microwave field. The profile of T_e remains largely uniform across all three investigated gas pressures, reinforcing the similarity in behavior between the octupole and decapole configurations.

In the case of the duo-decapole configuration, the electron density (N_e) is observed to be lower compared to the octupole, which demonstrates a significantly higher electron density. This variation highlights the influence of the multicusp configuration on plasma parameters, with the octupole providing a more favorable environment for achieving higher plasma density. The difference in N_e between the two configurations can be attributed to the differences in magnetic field geometry and the extent of confinement provided by the respective cusp arrangements. Additionally, the electron temperature T_e exhibits a distinct radial dependence in the duo-decapole configuration, decreasing progressively with increasing radial distance from the multicusp

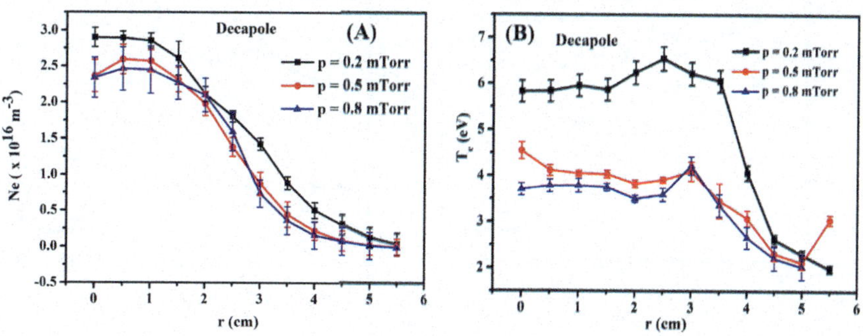

Fig. 6.3 Radial profile of **a** electron density, **b** electron temperature at three different pressures inside the decapole at microwave power of 180 W (reproduced with permission from author [Krishn Pal Singh, Optical properties of atomically heterogeneous systems created by microwave plasma-based low energy ion beams, Ph.D thesis, IIT Kanpur, India, 2023])

axis. The maximum T_e is observed at approximately 3.8 cm from the multicusp center, which is close to the location of electron cyclotron resonance (ECR), positioned at around 4.1 cm. This proximity to the ECR location suggests that the resonant absorption of microwave energy is a key factor in determining the peak T_e, as electrons gain maximum energy in this region before losing it through collisions or other mechanisms. The behavior of T_e in the duo-decapole configuration, as a function of varying neutral gas pressure, follows a trend similar to that observed in the octupole and decapole configurations. As illustrated in Fig. 6.4b, T_e remains relatively stable across the three examined gas pressures, but its magnitude decreases as the gas pressure increases. At lower gas pressures, T_e is higher, reflecting reduced collisional damping that allows electrons to retain more energy from the microwave field. Conversely, higher pressures lead to increased collisions between electrons and neutral atoms, dissipating electron energy and reducing T_e.

In comparison with the other configurations, the duo-decapole shows a lower overall T_e, aligning with its reduced N_e. On the other hand, the octupole exhibits the highest T_e and N_e values, underscoring its superior capability for plasma generation and energy transfer efficiency. This distinction highlights the octupole as the most effective multicusp configuration for applications requiring high plasma density and electron temperature, while the duo-decapole demonstrates relatively lower performance in these aspects.

In conclusion, the octupole configuration exhibits the highest values of electron density N_e and electron temperature T_e, whereas the duo-decapole configuration demonstrates the lowest values. Both T_e and N_e show a consistent decline with increasing neutral gas pressure across all multicusp setups. For T_e the decrease is particularly steep as the pressure rises, but it stabilizes at approximately 0.3 mTorr, remaining nearly constant beyond this point in all configurations. At elevated pressures, the electron temperature becomes comparable across the different multicusp geometries, indicating reduced sensitivity to the specific magnetic field arrangements.

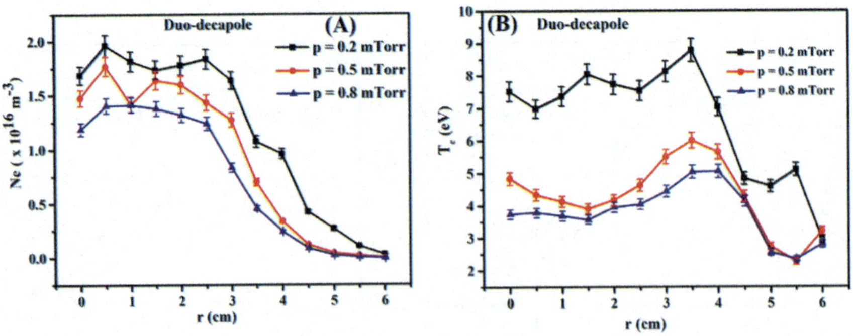

Fig. 6.4 Radial profile of **a** electron density, **b** electron temperature at three different pressures inside the duo-decapole at microwave power of 180 W (reproduced with permission from author [Krishn Pal Singh, Optical properties of atomically heterogeneous systems created by microwave plasma-based low energy ion beams, Ph.D thesis, IIT Kanpur, India, 2023])

The rate of change in T_e with pressure is most pronounced within the pressure range of 0.1–0.3 mTorr for each multicusp configuration, marking this interval as critical for controlling plasma characteristics. In contrast, N_e reaches its maximum at approximately 0.3 mTorr, followed by a decline that levels off at higher pressures. This trend suggests a saturation of plasma density under high-pressure conditions. Overall, the octupole configuration demonstrates superior performance in terms of maintaining higher plasma density and temperature, particularly at lower pressures, making it a preferred choice for applications requiring optimized plasma parameters. These findings highlight the influence of magnetic field geometry and gas pressure on plasma behavior, offering valuable insights for tailoring plasma sources to specific experimental or practical needs.

6.2 Ion-Flux Measurement of the Extracted Beam as a Function of Discharge Pressure and Microwave Power

Figure 6.5a represents the magnitude of ion-flux extracted from multicusp plasma at four different microwave powers, i.e., 180 W, 216 W, 252 W, and 288 W with varying gas pressure from 1×10^{-4} to 8×10^{-4} Torr in steps of 0.5×10^{-4} Torr [5]. The flux of Ar-ion reduces drastically up to 3.5×10^{-4} Torr, with the increase in gas pressure at a fixed microwave power, although the variation of decreasing flux beyond 3.5×10^{-4} Torr is slow. Further, the variation of Ar-ion-flux with microwave power is also recorded for three distinct gas pressures and presented in Fig. 6.5b. The extracted ion-flux increases linearly with increase in microwave power up to 234 W, beyond which an almost saturation behavior with minimal variation is observed. The maximum flux which can be obtained for Ar-ion is approximately 6×10^{14} cm^{-2} s^{-1}, which can be fine-tuned with above-mentioned parameters, i.e., microwave power and pressure of the gas.

Similarly, Ne, Ar, and Kr ions are also extracted through an electrode with a 1 cm aperture and the corresponding alteration in ion flux for Ne, Ar, and Kr ions with microwave power and pressure of the gas is depicted in the upcoming figures.

The change in Ne and Ar ion flux with gas pressure (from 1×10^{-4} to 6×10^{-4} Torr) and ion energy (from 0.5 to 2.5 kV) at constant microwave power is presented in Figs. 6.6 and 6.7, respectively. A decrease in ion-flux with increase in gas pressure is also observed here, although at higher extraction voltage, the ion flux is high. Further, the measurement of ion flux for Ar and Kr ions was conducted with varying applied voltage in a constant gas pressure of 1.2×10^{-4} Torr. The ion flux was recorded by changing the microwave power from 180 W and 270 W in steps of 18 W. Additionally, during Ne ion-flux measurements, the gas pressure is kept constant at constant 3.6×10^{-4} Torr. In a nutshell, the measured ion flux is higher at increased value of microwave power as well as extraction voltage (i.e., ion energy).

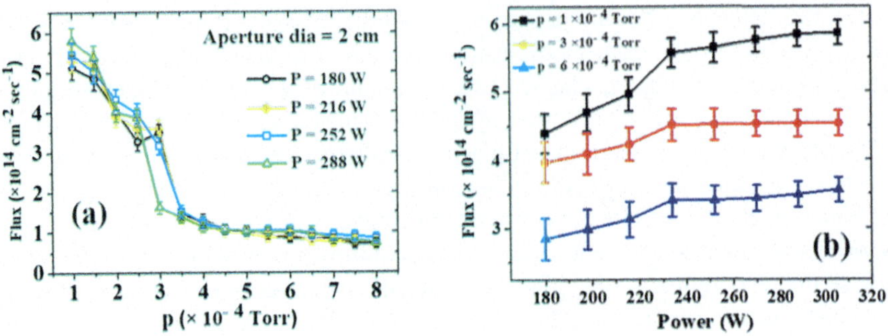

Fig. 6.5 Argon ion beam flux: **a** for fixed microwave power with varying pressure, 180 W (black open circle), 216 W (red solid diamond), 252 W (blue open square), and 288 W (green open triangle), **b** for fixed pressure with varying microwave powers from 180 W to 306 W in intervals of 18 W (reproduced with permission from author [Krishn Pal Singh, Optical properties of atomically heterogeneous systems created by microwave plasma-based low energy ion beams, Ph.D thesis, IIT Kanpur, India, 2023])

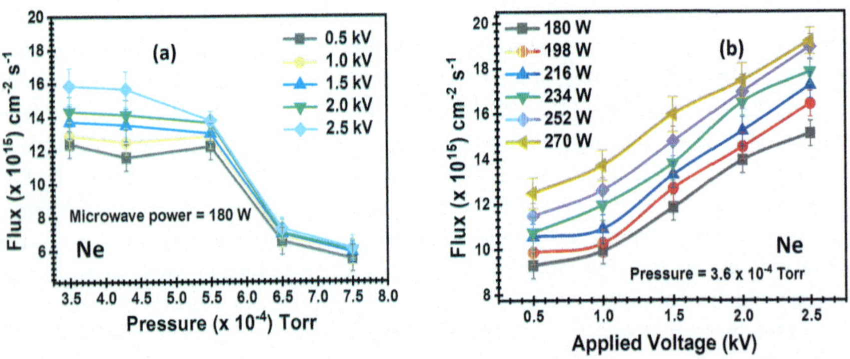

Fig. 6.6 Neon ion beam flux: **a** for fixed microwave power with varying extraction voltage, **b** for fixed pressure of 3.6×10^{-4} Torr with varying microwave powers from 180 W to 270 W in intervals of 18 W (reproduced with permission from author [Jayashree Majumdar, Field emission properties of atomically heterogeneous metallic microstructures created by microwave plasma generated low energy ion beams, Ph.D. thesis, IIT Kanpur, India, 2025])

6.3 Plasma Ion Saturation Current and Beam Uniformity

The radial and axial distributions of plasma density within the multicusp (MC) configuration (illustrated below) were evaluated using a Langmuir probe to determine the optimal conditions for achieving high ion flux. To carry out these measurements, a planar Langmuir probe with a stainless-steel tip measuring 4 mm in diameter was employed.

In general, the ion saturation current drawn from the plasma is given by [2].

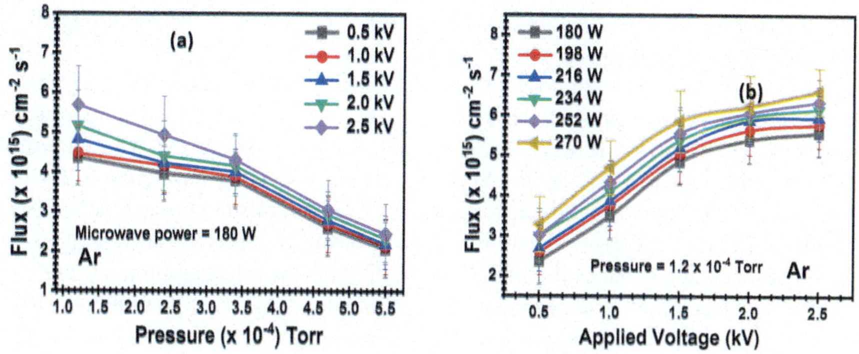

Fig. 6.7 Argon ion beam flux: **a** for fixed microwave power of 180 W and with varying extraction voltage, **b** for fixed pressure of 1.2×10^{-4} Torr with varying microwave powers from 180 W to 270 W in intervals of 18 W (reproduced with permission from author [Jayashree Majumdar, Field emission properties of atomically heterogeneous metallic microstructures created by microwave plasma generated low energy ion beams, Ph.D thesis, IIT Kanpur, India, 2025])

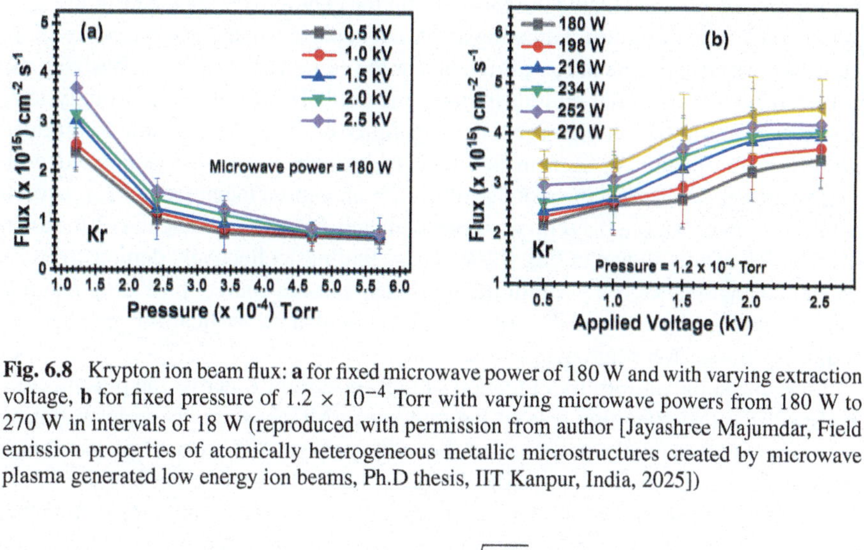

Fig. 6.8 Krypton ion beam flux: **a** for fixed microwave power of 180 W and with varying extraction voltage, **b** for fixed pressure of 1.2×10^{-4} Torr with varying microwave powers from 180 W to 270 W in intervals of 18 W (reproduced with permission from author [Jayashree Majumdar, Field emission properties of atomically heterogeneous metallic microstructures created by microwave plasma generated low energy ion beams, Ph.D thesis, IIT Kanpur, India, 2025])

$$I_s = 0.6 A n_i q \sqrt{\frac{\kappa T_e}{M}}. \tag{6.3}$$

Here the parameters A, n_i, q, κ, T_e, and M are the area of Langmuir probe, plasma (ion) density, charge of ion, Boltzmann's constant, electron temperature, and mass of ion, respectively. This equation offers information about maximum ion current that can be extracted from a plasma source. The Langmuir probe was operated by applying a high negative bias voltage of − 70 V to measure the ion saturation current I_s. Using this setup, the corresponding axial and radial current profiles within the multi-cusp (MC) configuration were systematically recorded. These measurements provide critical insights into the plasma density distribution, aiding in the optimization of operational conditions for enhanced ion flux.

The variation of ion saturation current (I_S) from an Argon plasma throughout the axis of the multicusp is presented in Fig. 6.9a. The variation is recorded at microwave power of 270 W and two different magnitudes of gas pressure [6]. The measurements commenced at a distance of 5 cm from the center of the multicusp (MC) and progressed inward toward the plasma core. To simulate the conditions encountered during ion beam extraction through a plasma electrode aperture, a dummy plate was positioned at the end of the MC. Outside the MC, the ion saturation current was observed to be negligible, as anticipated, but it increased significantly within the MC, reaching a saturation level approximately 4 cm inward from the starting point. The peak value of I_S, recorded near the center of the MC, was 1.6 mA. However, a slight decrease in I_s was observed with increase in gas pressure, indicating a pressure-dependent behavior. Figure 6.9b illustrates the radial variation of I_s at two gas pressures, 0.35 and 0.45 mTorr, under the same microwave power of 270 W. The measurements spanned the entire diameter of the MC (8.6 cm), beginning at one edge and extending to the other, with the probe positioned at an axial distance of approximately 0.4 cm outside the MC. Within the central 6 cm diameter of the MC, I_s remained nearly uniform, reflecting a consistent plasma density in the core region. However, a sharp decline in I_s was noted near the edges of the MC, emphasizing the confinement effects of the multicusp magnetic field. The observed uniformity in I_s across the central region highlights the capability of the plasma source to produce a uniform ion beam. As seen in the axial measurements, I_s also decreased with increasing gas pressure in the radial profile. The maximum magnitude of I_s recorded at the MC center, for ($z \sim 0.4$) cm, was approximately 0.4 mA, consistent with the axial distribution results shown in Fig. 6.10a. These findings collectively demonstrate the spatial characteristics of I_s within the multicusp plasma source, providing insights into the uniformity and stability of the plasma, which are critical for applications requiring precise ion beam generation.

To evaluate the uniformity of the ion saturation current I_s across the plasma electrode (PE) at the extraction side of the multicusp (MC) system, the polar variation of I_s was systematically investigated. For this purpose, "L"-shaped planar Langmuir probes were employed, specifically fabricated in three different lengths 0.5, 1, and 1.5 cm. These probes were rotated through a full 360° in angular increments of 22.5°, allowing for comprehensive measurements of I_s across the polar plane. Figure 6.10a and b depicts the angular distribution of I_s recorded at the edge of the MC confinement system. The experiments were conducted at two distinct microwave power levels, 270 W and 360 W, while maintaining the gas pressure at 0.25 mTorr. A color scale adjacent to the plots illustrates the magnitude of I_S, revealing a relatively uniform distribution of I_s over the measured area. At the center of the MC, I_s values were significantly higher, measured at approximately 0.6 mA for the 270 W power setting and 0.72 mA for the 360 W setting. This increase in I_s at higher microwave power is indicated by a darker region in the center of the plot, signifying enhanced ion current. These measurements were performed using the three Langmuir probes, each with an identical probe area, ensuring uniform probe characteristics and minimizing

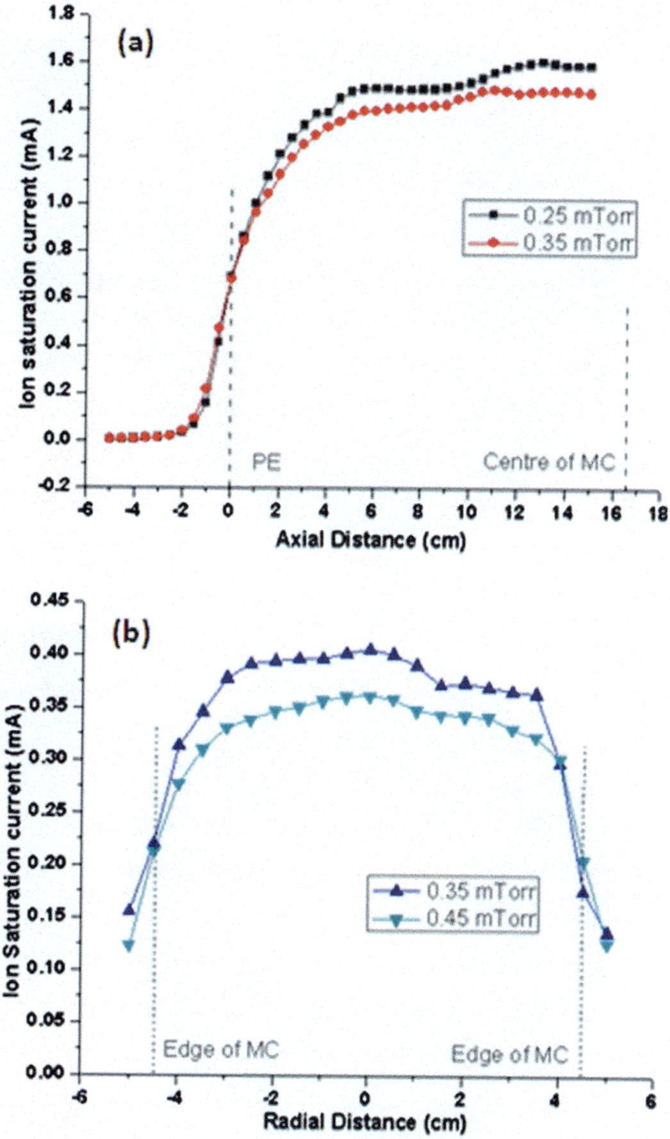

Fig. 6.9 **a** Axial and **b** radial variation of ion saturation current from an Argon plasma for microwave power 270 W (reprinted from [Abhishek Chowdhury and Sudeep Bhattacharjee, AIP Advances **1**, 042150 (2011)], with the permission of AIP)

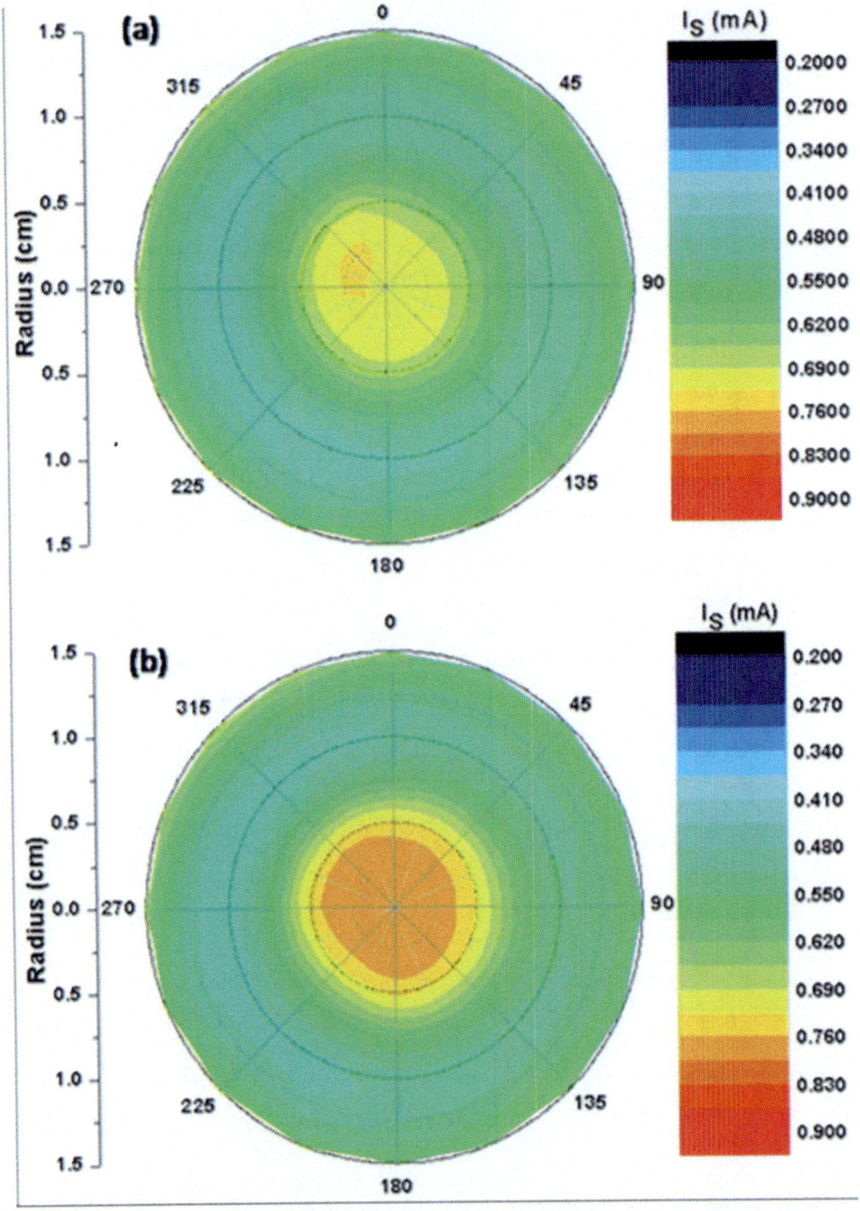

Fig. 6.10 Angular profile of ion saturation current from an Argon plasma at the end of MC for **a** microwave power 270 W and pressure 0.25 mTorr and **b** microwave power 360 W and pressure 0.25 mTorr (reprinted from [Abhishek Chowdhury and Sudeep Bhattacharjee, AIP Advances **1**, 042150 (2011)], with the permission of AIP)

measurement discrepancies. To simulate the conditions of the plasma electrode (PE) during these polar variation experiments, a dummy plate was utilized. This configuration differed from the setup used for the radial variation measurements shown in Fig. 6.9b, where no dummy plate was present. This difference in experimental setup likely accounts for the density variations observed in the central region during the polar variation study. The presence of the dummy plate may have caused microwave power reflections off the end plate, locally amplifying the wave power density and thereby influencing the I_s measurements. The ion density in the central region of the MC was estimated to be on the order of (10^{11}) ions per cubic centimeter, consistent with the elevated I_s values measured in this region. This high ion density underscores the efficiency of the MC confinement system in maintaining a dense and uniform plasma, which is critical for applications requiring precise ion beam extraction. The results provide valuable insights into the angular uniformity and central ion density characteristics of the MC system, contributing to its optimization for advanced plasma-based applications.

6.4 Sample Preparation Techniques for Ion Beam Irradiation

The atomically heterogeneous system was fabricated by subjecting metallic thin films, deposited on glass substrates, to ion bombardment. Metallic thin films were chosen for this study due to their critical applications in metal plating for satellites and spacecraft. These films serve as protective layers for spacecraft exteriors and antennas, shielding them from the intense light, heat radiation, and other harsh conditions encountered in the outer regions of Earth's atmosphere. Metals like gold (Au), silver (Ag), and aluminum (Al) are particularly suitable as UV-VIS-NIR reflectors, as they efficiently reflect sunlight and heat radiation, making them indispensable for aerospace applications. Furthermore, since aluminum is widely used as the primary structural material in spacecraft, ion irradiation of aluminum films effectively simulates the environmental conditions experienced by spacecraft when exposed to low- and high-energy particles in space. To conduct a comprehensive analysis, metallic thin films of gold, silver, aluminum, and copper (Cu)—a cost-effective alternative—were selected for the study. These films were deposited onto highly transparent glass substrates to ensure compatibility with optical characterization techniques. The preparation process began with cleaning the glass substrates to remove contaminants. The substrates were ultrasonically cleaned in acetone and isopropyl alcohol for 30 minutes each, followed by a thorough rinse with deionized (DI) water to eliminate any residual carbon contamination. The deposition of metallic thin films was carried out using a thermal evaporation system. This method involved evaporating the metals in a high vacuum, allowing them to condense onto the glass substrates. The system was equipped with an integrated quartz crystal thickness monitor, enabling precise control and monitoring of the film thickness during deposition. Films of copper, silver, gold, and aluminum with a nominal thickness of approximately 200 nm were successfully deposited. To ensure accuracy, the thickness of the deposited films was verified using a stylus profilometer. Before measuring the actual samples, the profilometer was calibrated using a standard sample of known thickness. Any discrepancies between the measured and stated thickness of the standard sample were documented, and subsequent measurements for other samples were adjusted to maintain precision. The thickness variation across the metallic thin films was systematically analyzed, with Fig. 6.11 [7] presenting the results for a representative selection of silver thin films. For each sample, three to four thickness measurements were conducted at different locations, and the average value was reported. The error bars in the figure represent the maximum and minimum deviations from the average thickness. After deposition, the metallic thin films were carefully stored in a vacuum desiccator to prevent atmospheric contamination and oxidation, preserving their structural integrity and ensuring reliable experimental results. The successful fabrication and characterization of these atomically heterogeneous surfaces, formed through ion bombardment on metallic thin films, provide a robust platform for investigating their optical, thermal, and surface properties. This study highlights the importance of understanding how such films interact with extreme environmental conditions, advancing their potential applications in aerospace and related fields.

The field emission measurements of atomically heterogeneous systems were conducted at ambient temperature without any external heating applied to the system. The copper sheets of dimensions $1 \times 1 \times 0.44$ cm^3, purity: > 99.999% were immersed in a 2 mol/L nitric acid (HNO_3) solution for 45 s to remove any surface hydroxide layers from the sample effectively. Following this, the samples were cleaned ultrasonically with acetone and ethanol to ensure the complete removal of surface contaminants. These cleaned samples were mounted on the collector plate

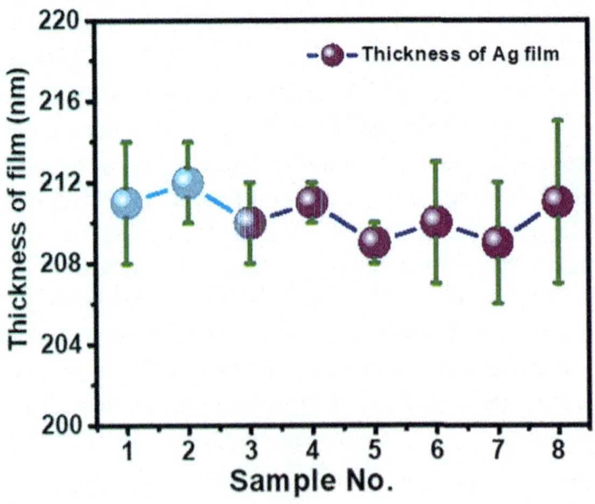

Fig. 6.11 Variation of thickness of the thin Ag films over a few number of samples deposited in similar condition (reprinted with permission from [7] © Optica Publishing Group)

for subsequent experiments. Both pristine and ion-irradiated samples were stored in isolated compartments in a vacuum desiccator (10^{-2} Torr) to avoid oxidation, moisture absorption, and environmental contamination (carbon) ensuring the integrity of their surface properties. Before imaging the surface in atomic scale via transmission electron microscopy (TEM), the copper substrates were reduced to circular disks of a diameter of 3 mm and thickness of 100 μm, mechanically. These disks were then subjected to a precise dimpling process, which involved thinning the central region, followed by ion milling. This additional thinning step reduced the central thickness to less than 40 μm, creating a suitable region for TEM characterization. The approximately 40-μm-thick central zone provided the necessary transparency for detailed TEM investigations, allowing for high-resolution analysis of the structure of the material as well as surface morphology.

Questions

1. How does a multicusp confine a plasma? How is the radial and axial plasma confinement achieved in a multicusp?
2. How can a Langmuir probe be employed to determine the plasma (ion) density and the electron temperature of a plasma? What are the formulae used to deduce these plasma parameters?
3. What is the difference between ion beam flux and fluence? How can fluence be determined from the flux? What are the units of flux and fluence?
4. What is meant by electron cyclotron resonance (ECR)? Calculate the magnetic field for ECR for waves of 2.45 GHz.
5. Study about the mechanism of magnetic mirror motion and loss cone for particles confined by a magnetic mirror. Can you estimate the percentage loss of particles through the magnetic cusp regions in a multicusp? (Please see the following reference: Sudeep Bhattacharjee and Hiroshi Amemiya, Production of microwave plasma in narrow cross-sectional tubes: Effect of the shape of cross section, Review of Scientific Instruments, 70, 3332, (1999)).
6. Please have a look at the plot of the magnetic field for different types of multicusp as shown in Fig. 6.1e, why is a larger flatter region observed for the duodecapole?
7. Have a look at Figs. 6.2 and 6.3 on the electron temperature profiles in a multicusp, can you think of a reason for the rise in electron temperature near the plasma periphery before it starts to decrease? [Hint: Please look up upper hybrid resonance in a plasma and see the following article: Debaprasad Sahu and Sudeep Bhattacharjee, Utilizing upper hybrid resonance for high density plasma production and negative ion generation in a downstream region, *J. Appl. Phys.* 112, 063304 (2012)).
8. Have a look at Figs. 6.7 and 6.8 on the ion flux for Ar and Kr, can you explain why the flux for Kr is lower than that of Ar at the same condition of microwave power and discharge pressure?

9. A cylindrical probe of radius $r_p = 0.5$ mm and length $L_p = 10$ mm is immersed in a collision-less, stationary, singly ionized argon plasma with an ion density of $n_i = 5 \times 10^{16}$ m^{-3}. The plasma has an electron temperature of $T_e = 3$ eV, and the ions are assumed to enter the sheath at the Bohm velocity. Given that, $m_i = 40$ amu, $e = 1.6 \times 10^{-19}$ C, $m_e = 9.11 \times 10^{-31}$ kg, and $m_p = 1.67 \times 10^{-27}$ kg (a) calculate the ion saturation current I_s and (b) how would the ion saturation current change if the plasma density increased to 1×10^{17} m^{-3}.
10. A thin film of indium tin oxide (ITO) is deposited on a glass substrate. A four-point probe measurement is used to determine its sheet resistance (R_s), and a separate profilometer measurement provides the thickness (t). If the sheet resistance $R_s = 20$ Ω/nm^2, thickness of the thin film is $t = 100$ nm and the resistivity $\rho = R_s \times t$, then (a) calculate the resistivity of the thin film, (b) if the thickness measurement has an uncertainty of ± 5 nm, determine the possible range of resistivity values, (c) if the sheet resistance remains the same but the film thickness is doubled, how does the resistivity change?

References

1. F.F. Chen et al., *Introduction to Plasma Physics and Controlled Fusion*, vol 1 (Springer, 1984)
2. S. Bhattacharjee, H. Amemiya, Microwave plasma in a multicusp circular waveguide with a dimension below cutoff. Jpn. J. Appl. Phys. **37**, 5742 (1998)
3. I. Dey, S. Bhattacharjee, Experimental investigation of standing wave interactions with a magnetized plasma in a minimum-B field. Phys. Plasmas **15**(12), 123502 (2008)
4. D. Sahu, S. Pandey, J. Aneja, S. Bhattacharjee, Negative ion rich plasmas in continuous and pulsed wave modes in a minimum-B magnetic field. Phys. Plasmas **19**(12), 123517 (2012)
5. K.P. Singh, Optical properties of atomically heterogeneous systems created by microwave plasma-based low energy ion beams. Ph.D thesis, IIT Kanpur, Chap. 2, August 2023
6. A. Chowdhury, S. Bhattacharjee, Localized subsurface modification of materials using micro-low-energy multiple ion beamlets. AIP Adv. **1**, 042150 (2011)
7. K.P. Singh, J. Majumdar, S. Bhattacharjee, Tuning optical properties of atomically heterogeneous systems created by plasma-based low-energy ion beams. Appl. Opt. **59**, 4507 (2020)

Chapter 7
Case Study 1: Modification of Electrical Properties: Theoretical Background and Experimental Procedures

This chapter delves into the theoretical framework and initial experimental investigations pertaining to the modification of electrical properties of materials through ion beam techniques. By exploring the underlying principles and the effects of ion irradiation on material systems, this chapter aims to establish a comprehensive understanding of the theoretical aspects driving changes in electrical behavior. The discussion includes both theoretical insights and experimental observations, providing a foundation for advancing the application of ion beam methods in tailoring material properties. In the consecutive chapter, the experimental results are presented in a systematic way that reveals the change in electrical properties of ion beam modified materials. The low energy ion beams (LEIBs) (0–3 keV)-induced material modification has garned significant attention over several years and is an area of dynamic research [1–9]. The growing interest is driven by the unique ability of LEIBs to induce surface modifications, manipulate material properties at the atomic scale, and enable advancements in fields such as nanotechnology, materials science, and device fabrication. LEIBs within the energy range of 0–1 keV are utilized to systematically study the modifications in the electrical sheet resistance (R_s) of metallic thin films. This controlled approach involves varying the ionic species and beam energy to achieve precise tuning of material properties. Broad ion beams are employed to ensure uniform irradiation across the sample surface, thereby enabling consistent modification and enhancing the sensitivity required for accurate measurement of R_s changes. The potential for inducing localized modifications at micron scales is explored, leveraging the predictions derived from experimental results. The data reveal a significant increase in the R_s of metallic thin films, correlating with both ion beam energy and the type of ionic species used. For instance, R_s exhibits a notable 50% rise at 1 keV when irradiated with argon (Ar) ions, and an even more pronounced ~ 90% increase is observed with krypton (Kr) ions at 0.6 keV. These experiments were conducted on 80 nm-thick copper films using 2 cm diameter ion beams. Atomic force microscopy (AFM) measurements confirm that ion-induced surface roughening is the primary mechanism driving these changes. Furthermore,

S. Bhattacharjee, *Atomically Heterogeneous Systems*,
https://doi.org/10.1007/978-981-95-9733-8_7

predictions regarding the changes in R_s induced by micro-beamlets are also analyzed, providing insights into the scalability and precision of this approach for localized applications.

7.1 Origin of Resistivity in Metals and the Matthiessen Rule

The electrical resistivity in metalic conductors arises due to the scattering of conduction electrons by various mechanisms. These include interactions with vibrating atoms in the crystal lattice, known as phonons, as well as collisions with impurities and structural imperfections within the material. Each of these factors contributes to the overall resistance, with phonon scattering typically dominating at higher temperatures, while impurity and defect scattering play a more significant role at lower temperatures. Each scattering process is associated with a distinct relaxation time, and Matthiessen's rule [10] allows for the determination of an effective relaxation time by combining these individual times, given by,

$$\frac{1}{\tau_{\text{avg}}} = \frac{1}{\tau_{\text{lattice}}} + \frac{1}{\tau_{\text{GB}}} + \frac{1}{\tau_{\text{defects}}} + \frac{1}{\tau_{\text{impurity}}} + \cdots \tag{7.1}$$

where τ_{lattice}, τ_{GB}, τ_{defects}, and τ_{impurity} are the relaxation times due to phonon-induced scattering, scattering induced by grain boundaries (GB), defects, and impurities respectively. If Eq. (7.1) is multiplied by $\frac{m_e}{ne^2\rho_o}$ we obtain,

$$\frac{\rho_{\text{avg}}}{\rho_o} = \frac{\rho_{\text{lattice}}}{\rho_o} + \frac{\rho_{\text{GB}}}{\rho_o} + \frac{\rho_{\text{defects}}}{\rho_o} + \frac{\rho_{\text{impurity}}}{\rho_o} + \cdots \tag{7.2}$$

where ρ_o and ρ_{avg} are the resistivity of the bulk metal and average resistivity of the thin film and ρ_{lattice}, ρ_{GB}, ρ_{defects}, and ρ_{impurity} are the individual resistivity contribution of the thin film due to effects of lattice, grain boundary, defects, and impurities, respectively. The mass, charge and number density of electrons are given by m_e, e, and n, respectively. In a thin polycrystalline film with a thickness on the order of the mean free path of the bulk metal, the dominant scattering mechanisms are surface scattering and grain boundary scattering.

7.2 Electron Transport in Single-Crystal and Poly-crystalline Films

According to Fuchs [11], the theory of electron transport for single-crystal films is articulated with a formula that relates thin-film resistivity to bulk resistivity, given by

$$\frac{\rho_{\text{lattice}}}{\rho_o} = \left[1 - \left(\frac{3}{2k}\right)(1-p)\int_d^{\infty}\left(\frac{1}{t^3} - \frac{1}{t^5}\right)\left(\frac{1-e^{-kt}}{1-pe^{-kt}}\right)dt\right]^{-1}, \tag{7.3}$$

where $k = d/\lambda_o$, p is the coefficient of specular reflection. $p = 1$ indicates all the scattering is specular and the resistivity is equal to that of the bulk material, whereas $p = 0$ means completely diffused scattering. Equation (7.3) represents the first term on the right hand side (RHS) of Eq. (7.2).

In their model for polycrystalline films, Mayadas and Shatzkes [12] approximated grain boundaries as δ-function potentials and introduced a formula to calculate the ratio of grain boundary resistivity to that of the bulk material, given by,

$$\frac{\rho_{\text{GB}}}{\rho_o} = \left[1 - \frac{3}{2}\alpha + 3\alpha^2 - 3\alpha^3 \ln\left(1 + \frac{1}{\alpha}\right)\right]^{-1}, \tag{7.4}$$

where $\alpha = \lambda_o/d \times (R/(1-R))$, R = grain boundary reflection coefficient. The resistivity of pure polycrystalline films can be assessed using Eq. (7.4), which represents the second term on the right-hand side of Eq. (7.2). This correlation is well-supported by experimental data, as detailed in the comparison of appendix of various experimental results with Eqs. (7.3) and (7.4). The third and fourth terms on the right-hand side of Eq. (7.2) account for changes in resistivity caused by defects and impurities introduced through ion beam irradiation. These terms are determined experimentally by analyzing the variation in sheet resistance (R_{RS}). To isolate the phonon contribution, the temperature dependence of R_S in the irradiated films is analyzed, enabling the determination of the change in R_{RS}.

7.3 Temperature Dependence of Resistivity of a Metal

In metals, the resistivity ρ as a function of temperature is given by the Bloch–Gruneisen formula [13] as,

$$\rho(T) = \rho_o + A\left(\frac{T}{\theta_D}\right)^5 \int_0^{\theta_D/T} \frac{x^5}{(e^x-1)(1-e^x)}dx, \tag{7.5}$$

where ρ_o is the residual resistivity, A is the prefactor of the Bloch–Gruneisen formula, and θ_D is the Debye temperature of the metal. The property θ_D is intrinsic to the material and results from electron–phonon interactions. As fluence and mass increase, θ_D shows a reduction from its pristine value. Here θ_D is given by

$$\theta_D = \frac{\hbar\omega_D}{\kappa_B} = \frac{\hbar v_s}{\kappa_B}\left(\frac{6\pi^2 N}{V}\right)^{1/3}, \tag{7.6}$$

where $\hbar$, $\kappa_{B,}$ and v_s are reduced Planck's constant, Boltzmann constant, and velocity of sound in the sample, respectively, and N is the total number of phonons in the volume V and ω_D is the cut-off frequency associated with θ_D. At frequencies higher than ω_D, the lattice becomes independent of the incoming vibration.

During irradiation, the volume of the thin film is assumed to remain constant, with the number density remaining relatively constant for a particular sample. In this scenario, the decrease in the diffusion coefficient D is primarily caused by a reduction in the sound velocity v_s, which is directly proportional to D. The sound velocity v_s in solids is dependent on the bulk modulus E and the material density δ, as given by the relation $v_s = \sqrt{E/\delta}$. To estimate the density change in the film post-irradiation, it is assumed that the volume of the film does not alter. For instance, an 80-nm thick silver film with a 1 cm^2 area has a mass of 8.44×10^{-5} g. Irradiating this film with Kr^+ (mass number 84) at a maximum fluence of 1.38×10^{16} ions cm^{-2} adds an extra mass of 1.93×10^{-6} g, leading to a increase in density by 2.3%. This represents the highest expected change, calculated for the heaviest ion Kr at maximum fluence; for other mass and fluence scenarios, the density change will be less than 2.3%. Therefore, the primary factor contributing to the decrease in v_s is the elastic modulus E in the sample, with the density change playing a minimal role.

7.4 Experimental Details

The investigation utilizes a microwave-generated multicusp plasma ion source to generate the ion beam, as detailed in prior references [14–16]. The beam fluence can be adjusted within the range of approximately 4.4×10^{15} to 7×10^{15} ions per square centimeter, as mentioned earlier. Detailed descriptions of the ion source and irradiation setup are available in the referenced literature [14–16] and earlier chapters. Ion beams with a diameter of about 20 mm are employed to uniformly irradiate samples measuring 10 mm × 10 mm with a thickness of 80–200 nm. Metallic thin films are fabricated on glass substrates via thermal evaporation at a slow deposition rate of roughly 0.4 Å per second to ensure uniformity. During irradiation, the thin film is positioned 2 mm from the plasma electrode, and the process lasts for 2 s. Longer irradiation durations were found to either etch away or blacken the thin film due to excessive heat. The van der Pauw method [17, 18], described in earlier chapters, is used to measure the sheet resistance (R_S). The resistivity (ρ) is then calculated using the formula $\rho = (R_S \times d)$. Two sets of experiments are conducted: in the first, thin films are irradiated with argon ion beams at varying fluences of 0.3 keV energy; in the second, films are exposed to different ionic species (H_2, N_2, Ne, Ar, Kr) with mass numbers ranging from 1 to 84 at a fixed fluence of 9.6×10^{15} ions per square centimeter. Experimentally, the temperature dependence of R_S in the pristine film is expected to correspond with the behavior of the lattice resistivity ($\rho_{lattice}$). Temperature variations of R_S of irradiated films are measured to subtract the phonon contribution and isolate the sum of the grain boundary resistivity

(ρ_{GB}), defect resistivity ($\rho_{defects}$), and impurity resistivity ($\rho_{impurity}$). The residual resistance ratio (R_{RR}), defined as the ratio of R_S at room temperature (295 K) to R_S at low temperature (15 K), was calculated from low-temperature measurements. EDX spectroscopy was conducted on argon-irradiated samples to study the impact of impurities, involving cross-sectional analysis with line and map scans to determine the depth profile and elemental mapping within the samples. This provided insights into the extent of gaseous ion penetration in the metallic thin film. The experiments were conducted using p-type silicon wafers (110) approximately 250 μm thick as substrates for depositing the metallic thin films.

7.5 Preliminary Measurement of Change in Electrical Sheet Resistance

The irradiations were carried out by several projectile ions i.e., H^+, N^+, Ne^+, Ar^+, and Kr^+. The change in R_s in Fig. 7.1a exhibits the alteration in R_s of Cu films as a function of ion energy. The relative change in Rs i.e., ($\Delta R_s/R_s$) can be traced in the blue line of Fig. 7.1a. The clear signature of increasing sheet resistance with beam energy (from 0 to 1 keV) is obvious from the Figure. Further the variation of R_s with projectile ion mass for a fixed ion fluence is also depicted in Fig. 7.1b. The pre-bombardment magnitude of R_s of the films were almost negligible ($\sim 2\ \Omega$).

The alteration of sheet resistance is negligible during bombardment with lower projectile mass; however, the increase in projectile ion mass results in an increase in sheet resistance. The change in the RMS roughness of the surface can be employed to identify the reason of alteration of sheet resistance in the post-bombardment scenario. The increase in RMS surface roughness due to bombardment with higher projectile ions, as investigated by AFM, is shown in Fig. 7.2a–d. 3D AFM images of pristine and nitrogen, neon, and krypton ion-bombarded Cu films are presented in Fig. 7.2a and b–d. The magnitudes of RMS roughness of pristine and ion bombarded Cu films are listed in Fig. 7.2. The heavier projectile ions are mostly implanted near the surface or subsurface region with different depths from the top surface, resulting in an increment of R_s with projectile ion mass.

An experiment was conducted using unirradiated copper (Cu) films with varying surface areas in the range of 50–250 mm^2 and a thickness of 80 nm. The sheet resistances of these films were measured, as shown in Fig. 7.3. The data points can be approximately fitted with a straight line. From the linear relationship between sheet resistance (R_s) and film area, the R_s for a micrometer-sized area of 2500 μm^2 (approximately the area of the ion beamlets) was estimated to be 4.3 Ω/sq. The change in R_s upon beam irradiation is given by $\Delta R_s = (R_{si} - R_{so})$, where R_{si} and R_{so} (4.3 Ω/sq) are the sheet resistances after and before irradiation, respectively. The values of ΔR_s, which are expected to be constant, can be obtained from Fig. 7.3, and the predicted values of R_{si} for samples irradiated with micrometer-sized beamlets are

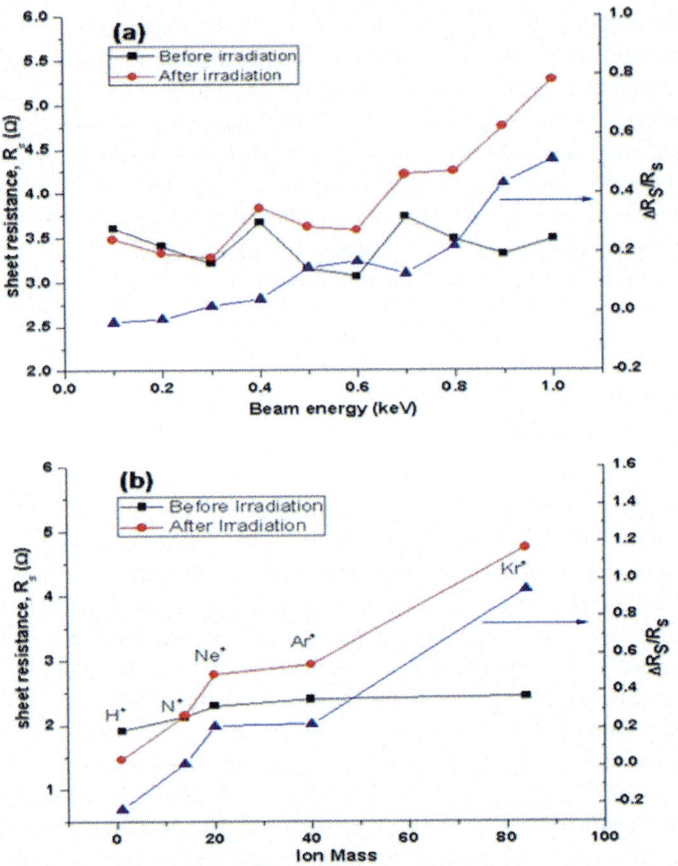

Fig. 7.1 Change in sheet resistance of Cu films upon ion irradiation with ions of **a** different energy and **b** different mass. In **a** Ar ion beams have been used. The error in measurement of the sheet resistance is ± 0.01 Ω. The sheet resistance indicated are in units of ohm/sq. Reprinted from [Abhishek Chowdhury and Sudeep Bhattacharjee, AIP Advances 1, 042150 (2011)], with the permission of AIP Publishing

provided in Fig. 7.3. The 3D AFM images of the ion irradiated thin films is shown in Fig. 7.4 and the scaling relation between resistivity and the area of the film can be obtained from Fig. 7.5.

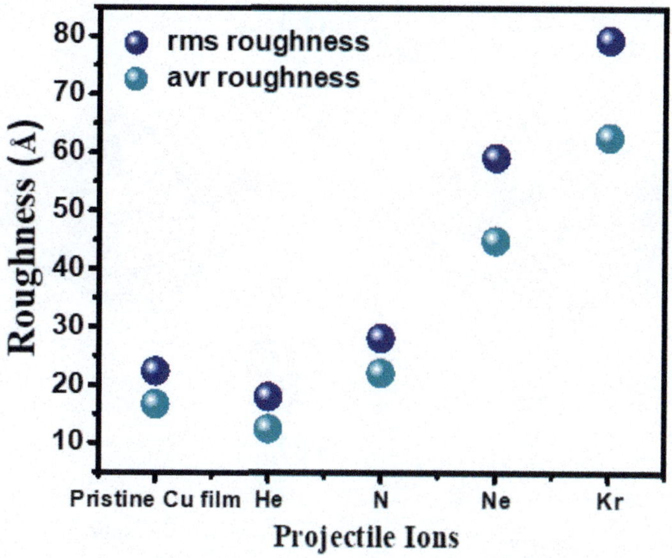

Fig. 7.2 Variation of magnitude of RMS roughness and average roughness of pristine and ion bombarded Cu films

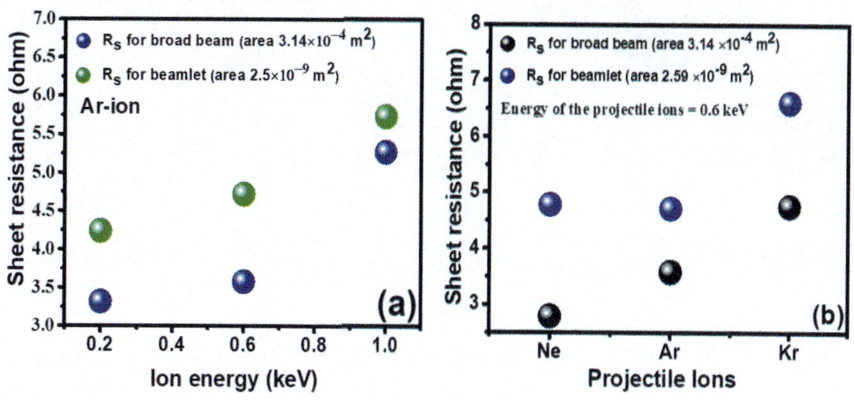

Fig. 7.3 Variation of **a** sheet resistance with energy of Ar-ion, **b** sheet resistance with projectile ions with energy 0.6 keV. The variations are recorded for broad beam and beamlets. The sheet resistance indicated are in units of ohm/sq

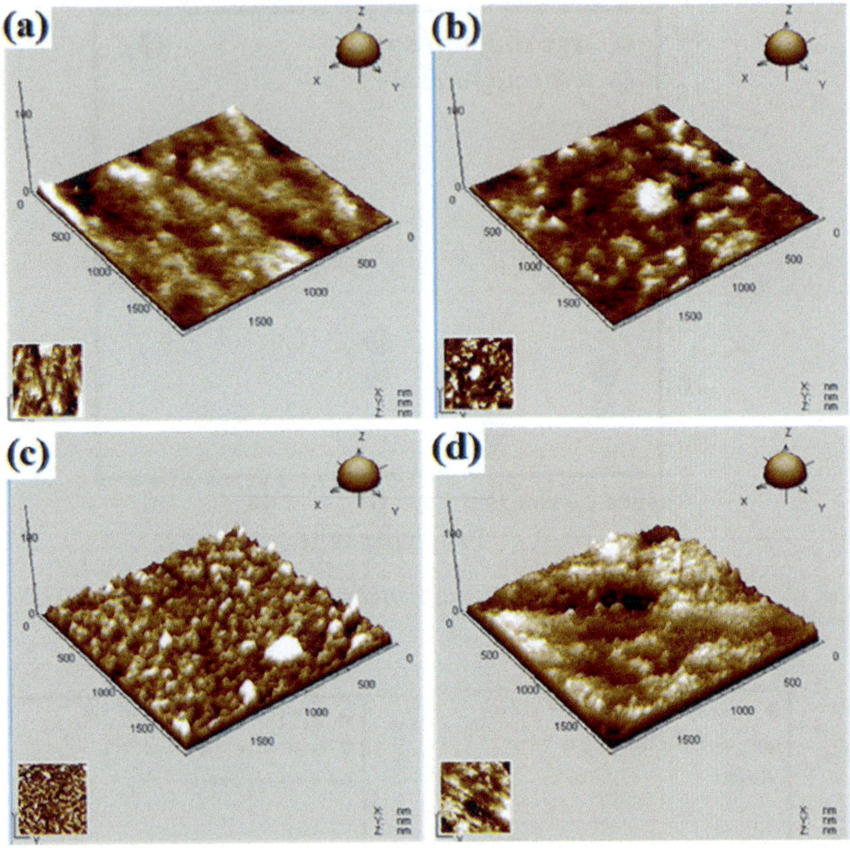

Fig. 7.4 3D AFM images of Cu films **a** as deposited and irradiated with **b** nitrogen, **c** neon, and **d** krypton ion. Reprinted from [Abhishek Chowdhury and Sudeep Bhattacharjee, AIP Advances 1, 042150 (2011)], with the permission of AIP Publishing

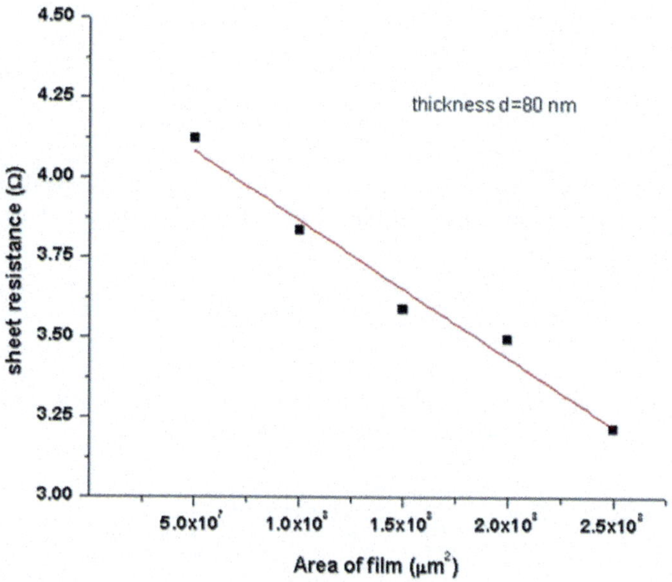

Fig. 7.5 Graph for obtaining a scaling relation between resistivity and the area of the film. The sheet resistance indicated are in units of ohm/sq. Reprinted from [Abhishek Chowdhury and Sudeep Bhattacharjee, AIP Advances 1, 042150 (2011)], with the permission of AIP Publishing

Appendix

The sheet resistance of single-crystal thin films is determined by examining electron transport as governed by the Boltzmann transport equation [10], given by:

$$\frac{\partial f}{\partial t} + \frac{\partial f}{\partial \vec{x}} \cdot \frac{\vec{p}}{m} + \frac{\partial f}{\partial \vec{p}} \cdot \vec{F} = \frac{\partial f}{\partial t} \mid \text{coll}, \tag{7.7}$$

which needs to be solved, where f represents the electron distribution function within the metallic thin film, $\vec{F}$ is the external force acting on the electrons, which can result from any applied electric field, $\vec{p}$ denotes the electron momentum, and $(\partial f/\partial t)$ represents the rate of change of the distribution function over time, which is zero at steady state. $(\partial f/\partial x)$ indicates the change in the distribution function with respect to position. The collisional term $(\partial f/\partial t)$ can be defined as,

$$\frac{\partial f}{\partial t} \mid \text{coll} = -\frac{(f - f_0)}{\tau} = -\frac{f_1}{\tau}, \tag{7.8}$$

where f_0 is the unperturbed electron distribution function in the absence of the external force (i.e., electric field); f_1 is the change in the unperturbed electron distribution function in time τ which is the electron relaxation time. At steady state, the above Eq. (7.7) reduces to

$$\frac{\partial f}{\partial \overrightarrow{x}} \cdot \frac{\overrightarrow{p}}{m} + \frac{\partial f}{\partial \overrightarrow{p}} \cdot \overrightarrow{F} = -\frac{(f - f_0)}{\tau}. \tag{7.9}$$

Assuming that the external electric field E_{ex} is applied in the x-direction, with the film lying in the x–y plane and its thickness along the z-axis, the equation simplifies to a scalar form given by:

$$V_z \frac{\partial f_1}{\partial z} + \frac{f_1}{\tau} = -\frac{eE_{ex}}{m} \frac{\partial f_0}{\partial V_x}, \tag{7.10}$$

where V_x and V_z are the components of electron velocity along the x and z directions, respectively. f_1 is obtained by solving Eq. (7.11) and is given by

$$f_1 = -\frac{eE_{ex}\tau}{m} \frac{\partial f_0}{\partial V_x} \left[1 + \phi \exp\left(-\frac{z}{\tau V_x} \right) \right], \tag{7.11}$$

where ϕ is an integration constant, to be determined from the boundary conditions of f_1. The current density can be obtained from $J = 2e(m/h)^3 \int (V_x f_1)\mathrm{dv}$. The resistivity of the system is defined as $\rho = E/J$.

For single-crystal films, the theory of surface scattering is given by Fuchs [11]. The ratio of thin-film resistivity to the bulk resistivity is given by the formula,

$$\frac{\rho_{\text{lattice}}}{\rho_o} = \left[1 - \left(\frac{3}{2k} \right) (1 - p) \int_d^{\infty} \left(\frac{1}{t^3} - \frac{1}{t^5} \right) \left(\frac{1 - e^{-kt}}{1 - pe^{-kt}} \right) \mathrm{d}t \right]^{-1}, \tag{7.12}$$

where $k = d/\lambda_o$, p is the coefficient of specular reflection. A value of $p = 1$ signifies that all scattering is specular, resulting in a resistivity equal to that of the bulk material, while $p = 0$ indicates completely diffused scattering. Figure 7.6 presents the experimental data for single-crystal silver films [19]. The data align with the theoretical prediction from Eq. (7.12) with $p = 0.5$. Therefore, Eq. (7.12) is suitable for evaluating the resistivity of pure single-crystal films, as it matches well with experimental results.

Thin metal films are often not composed of single crystals. Rather, they are more accurately described as arrays of randomly oriented polycrystalline films or grains. This polycrystalline structure introduces additional resistance due to scattering at the grain boundaries. To analyze the impact of these grain boundaries, Mayadas and Shatzkes [12] developed a model that treats the grain boundaries as delta-function potentials. The ratio of grain boundary resistivity to bulk resistivity is given by the following formula:

$$\frac{\rho_g}{\rho_o} = \left[1 - \frac{3}{2}\alpha + 3\alpha^2 - 3\alpha^3 \ln\left(1 + \frac{1}{\alpha} \right) \right]^{-1} \tag{7.13}$$

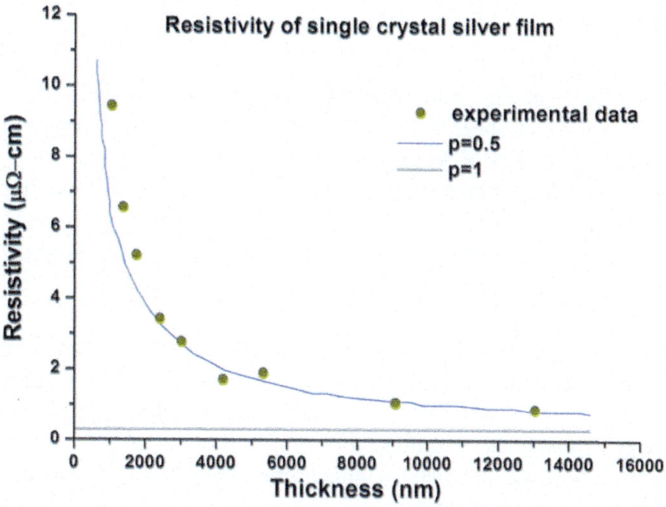

Fig. 7.6 Thickness-dependent resistivity for single-crystal silver film. The solid lines are the plot of Eq. (7.10) for different p. Reprinted from [Abhishek Chowdhury and Sudeep Bhattacharjee, J. Phys. D: Appl. Phys. 46 (2013) 435304)], with the permission of IOP Publishing

where $\alpha = \lambda_0/d \times R/(1 - R)$, where R = grain boundary reflection coefficient. Figure 7.7 illustrates the variation of resistivity with film thickness for a polycrystalline copper film. The experimental data are sourced from [20], with theoretical lines plotted for $R = 0.4$.

As a result, Eq. (7.13) may be employed to estimate the resistivity of pure polycrystalline films, as it reasonably matches experimental findings.

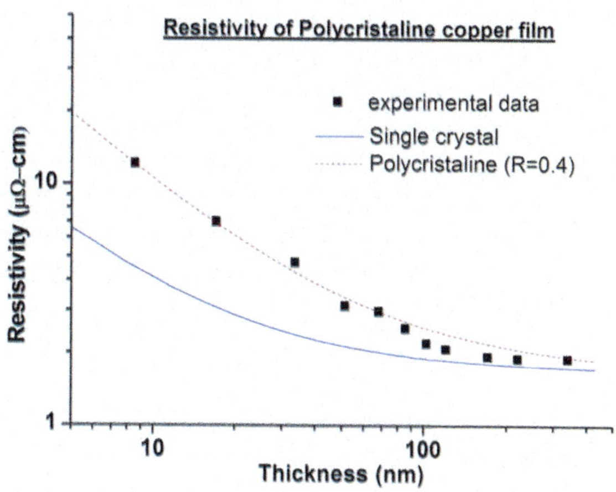

Fig. 7.7 Thickness-dependent resistivity for polycrystalline copper film. Reprinted from [Abhishek Chowdhury and Sudeep Bhattacharjee, J. Phys. D: Appl. Phys. 46 (2013) 435304)], with the permission of IOP Publishing

Questions

1. What are the typically energies of low energy ion beams (LEIB)?
2. What is electrical sheet resistance? How does it differ from resistance of conventional cylindrical wires that we measure?
3. What are the different factors attributing to the resistivity of conduction electrons in a metal?
4. What are the dominant electron scattering processes in polycrystalline thin films where the thickness of the film is of the order of the mean free path of the bulk metal?
5. What is the difference between single and polycrystalline films? From where does the additional resistivity arise in polycrystalline films? Can you give example of some materials whose polycrystalline films can be found? What are the typical size of the grains?
6. What is the formula that dictates temperature dependence of resistivity in materials? What is meant by Debye temperature? What is the possible cause of the decrease in Debye temperature in the material due to ion beam irradiation?
7. Read about the van der Pauw method of measuring the sheet resistance of metallic thin films. How does it compare with the conventional method of measuring the resitance of a wire?
8. What is the difference between the resistivity measurement at room temperature and at low temperature? What additional information may be obtained by undertaking the measurments at low temperature?
9. What are the typical values of sheet resistance of Cu metallic thin films used in the experiment? To what extent is the sheet resistance modified by employing ions of different gases? Can similar results be achieved by varying the ionic energy?
10. What is the significance of Debye temperature? Why is the Debye temperature an important parameter in understanding the thermal properties of solids, and how does it relate to the specific heat and lattice vibrations of a material at different temperatures?
11. A metal sample has an experimentally determined Bloch–Grüneisen temperature of 180 K from resistivity measurements. The material is known to have an empirical scaling factor $\alpha = 1.2$, α is a material-dependent factor, typically ranging from 1 to 1.5 for metals. The Bloch–Grüneisen temperature (Θ_{BG}) is related to the Debye temperature (Θ_D) as $\Theta_{BG} \approx \Theta_D/\alpha$. Estimate the Debye temperature.

References

1. S. Vauth, S.G. Mayr, Ion bombardment induced smoothing of amorphous metallic surfaces: Experiments versus computer simulations. Phys. Rev. B **77**, 155406 (2008)

2. F. Frost, A. Schindler, F. Bigl, Roughness evolution of ion sputtered rotating InP surfaces: pattern formation and scaling laws. Phys. Rev. Lett. **85**, 4116 (2000)
3. J.E. Gayone, R.G. Pregliasco, G.R. Gomez, E.A. Sanchez, O. Grizzi, Topographic and crystallographic characterization of a grazing-ion-bombarded GaAs (110) surface by time-of-flight ion-scattering spectrometry. Phys. Rev. B **56**, 4186 (1997)
4. C. Steiner, B. Bolliger, M. Erbudak, M. Hong, A.R. Kortan, J. Kwo, J.P. Mannaerts, Structural modifications of the Gd_2O_3 (100) films on GaAs (100). Phys. Rev. B **62**, R10614 (2000)
5. S. Macko, F. Frost, B. Ziberi, D. Forster, T. Michely, Is keV ion-induced pattern formation on Si (001) caused by metal impurities? Nanotechnology **21**, 085301 (2010)
6. B. Ziberi, F. Frost, T. Hoche, B. Rauschenbach, Ripple pattern formation on silicon surfaces by low-energy ion-beam erosion: experiment and theory. Phys. Rev. B **72**, 235310 (2005)
7. P.F.A. Alkemade, Propulsion of ripples on glass by ion bombardment. Phys. Rev. Lett. **96**, 107602 (2006)
8. H. Hansen, A. Redinger, S. Messlinger, G. Stoian, Y. Rosandi, H.M. Urbassek, U. Linke, T. Michely, Mechanisms of pattern formation in grazing-incidence ion bombardment of Pt(111). Phys. Rev. B **73**, 235414 (2006)
9. M.P. Ray, R.E. Lake, C.E. Sosolik, L.B. Thomsen, G. Nielsen, I. Chorkendorff, O. Hansen, Subsurface excitations in a metal. Phys. Rev. B **80**, 161405(R) (2009)
10. T.J. Coutts, *Electrical Conduction in Thin Metal Films* (Elsevier, Amsterdam, 1974)
11. K. Fuchs, The conductivity of thin metallic films according to the electron theory of metals. Proc. Camb. Phil. Soc. **34**, 100 (1938)
12. A.F. Mayadas, M. Shatzkes, Electrical-resistivity model for polycrystalline films: the case of arbitrary reflection at external surfaces. Phys. Rev. B **1**, 1382 (1970)
13. J.M. Ziman, *Electrons and Phonons: The Theory of Transport Phenomena in Solids* (Clarendon, Oxford, 1960)
14. J.V. Mathew, A. Chowdhury, S. Bhattacharjee, Subcutoff microwave driven plasma ion sources for multielemental focused ion beam systems. Rev. Sci. Instrum. **79**, 063504 (2008)
15. A. Chowdhury, S. Bhattacharjee, Micron-focused ion beamlets. J. Appl. Phys. **107**, 093307 (2010)
16. A. Chowdhury, S. Bhattacharjee, Localized subsurface modification of materials using micro-low-energy multiple ion beamlets. AIP Adv. **1**, 042150 (2011)
17. L.J. van der Pauw, A method of measuring specific resistivity and Hall effect of discs of arbitrary shape. Philips Res. Rep. **13**, 1 (1958)
18. G.T. Kim, J.G. Park, Y.W. Park, C. Muller-Schwanneke, M. Wagenhals, S. Roth, Nonswitching van der Pauw technique using two different modulating frequencies. Rev. Sci. Instrum. **70**, 2177 (1999)
19. D.C. Larson, B.T. Boiko, Electrical resistivity of thin epitaxially grown silver films. Appl. Phys. Lett. **5**, 155 (1964)
20. J.-W. Lim, K. Mimura, M. Isshiki, Thickness dependence of resistivity for Cu films deposited by ion beam deposition. Appl. Surf. Sci. **217**, 95 (2003)

Chapter 8
Case Study 2: Temperature Variation of Sheet Resistance in Metals for Pristine and Ion Beam Irradiated Samples

The chapter delves into an in-depth exploration of the effects of low-energy ion beam irradiation on the sheet resistance and Debye temperature of metallic thin films composed of Ag, Cu, and Al. Systematic experimental investigations are conducted to elucidate these effects, with the film thickness carefully selected such that the ratio of film thickness to the bulk mean free path (d/λ_0) ranges from 2 to 5, where d is the film thickness and λ_0 is the bulk mean free path. The primary objective of the study is to induce controlled defects and incorporate impurities at various depths within the films by fine-tuning the ion beam parameters, including ion energy, implantation angle, and the type of projectile ion species. These controlled modifications enable a detailed investigation of the subsequent changes in sheet resistance and Debye temperature. Ion irradiation experiments are conducted using both noble atomic ions (Ne, Ar, and Kr) and reactive molecular ions (H_2, N_2), with ionic mass numbers varying between 1 and 84. The ion fluence is maintained within a narrow range of 8.7×10^{15}–1.4×10^{16} ions cm^{-2}, to ensure precise control over defect generation and impurity incorporation. Residual sheet resistance (R_{RS}) measurements are carried out for both pristine and ion-modified metallic thin films over a temperature range of 300–15 K. The observed variations in R_{RS} and Debye temperature (θ_D) with ion fluence and projectile ion mass are analyzed, and an empirical formula is proposed to describe these relationships.

The substantial changes in R_{RS} are primarily attributed to the significant diffusion of impurities introduced through projectile ion implantation, as confirmed by energy-dispersive X-ray spectroscopy (EDS). The temperature-dependent sheet resistance data is further analyzed using the Bloch-Grüneisen model to extract the Debye temperature. A systematic decrease in θ_D with increase in ion beam fluence and projectile ion mass is observed, which is linked to changes in the bulk modulus of the thin films, as discussed in preceding chapters. During ion beam irradiation, the primary crystalline defects in metals include vacancies (vacant lattice sites), interstitial atoms of the host metal, and substitutional defects caused by the incorporation

S. Bhattacharjee, *Atomically Heterogeneous Systems*,
https://doi.org/10.1007/978-981-95-9733-8_8

of projectile ions. Although these point defects are introduced at nominal concentrations, they significantly influence the physical properties and processes in metals. Consequently, this investigation emphasizes the role of defect formation in governing the residual sheet resistance, providing critical insights into the interplay between ion-induced defects and the fundamental properties of metallic thin films.

The temperature dependence of sheet resistance (R_S) for both pristine and ion-irradiated thin films across varying ion beam fluences for three metals: (a) silver, (b) copper, and (c) aluminum is depicted in Fig. 8.1. The value of R_S of pristine silver at room temperature is approximately 0.2 Ω, corresponding to a film resistivity of 3.83×10^{-8} Ω m, which is approximately 2.38 times greater than the bulk resistivity of silver. This value is in close agreement with the theoretical resistivity for polycrystalline silver films (3.79×10^{-8} Ω m). Similarly, aluminum exhibits a room-temperature resistivity of 16.8×10^{-8} Ω m, which is about six times higher than its bulk resistivity of 2.74×10^{-8} Ω m. For copper, the film resistivity is measured at 15.09×10^{-8} Ω m, roughly eight times greater than the bulk resistivity of copper, which is 1.7×10^{-8} Ω m. In pristine films, the R_S changes nearly linearly with temperature between 300 and 70 K, but below 50 K, it deviates from this linear trend, as predicted by the Bloch-Grüneisen formula (see Fig. 8.2). This behavior is also observed in pristine copper and aluminum films, where deviations from linearity occur around 40–60 K. For irradiated films, the residual sheet resistance (R_{RS}) increases as ion beam fluence increases. A summary of these findings is provided in Table 8.1.

8.1 Bloch-Grüneisen Equation Fitting of the Experimental Data for Pristine Metal Films

Figure 8.2 presents the fitting of the Bloch-Grüneisen formula to the experimental data for pristine films of (a) silver, (b) copper, and (c) aluminum. The formula shows a good match with the experimental data for both copper and silver films. The root mean square (RMS) deviation of the data points from the formula is 0.0127×10^{-8} Ω m for copper and 0.0026×10^{-8} Ω m for silver. However, for aluminum, the deviation is slightly higher, at 0.0526×10^{-8} Ω m.

8.2 Temperature Variation of Sheet Resistance for Pristine and Various Projectile Ion Beam-Induced Metallic Thin Films

The temperature-dependent variations of R_S for both pristine and different projectile ion-irradiated metallic thin films: (a) silver, (b) copper, and (c) aluminum, are depicted in Fig. 8.3. The variation of R_S for pristine films is similar to that observed

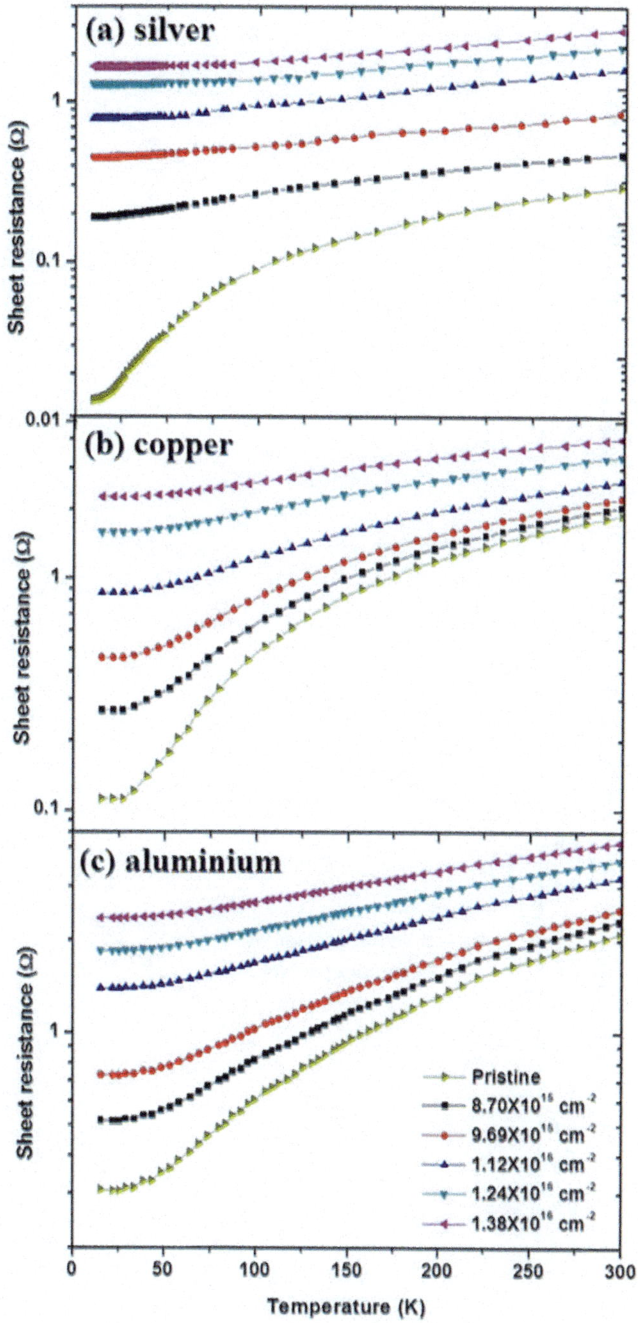

Fig. 8.1 Temperature variation of sheet resistance R_S of pristine and irradiated films with varying fluence for **a** silver, **b** copper, and **c** aluminum thin films. Reprinted from Abhishek Chowdhury and Sudeep Bhattacharjee, J. Phys. D: Appl. Phys. 46 (2013) 435304, with permission of IOP publishing

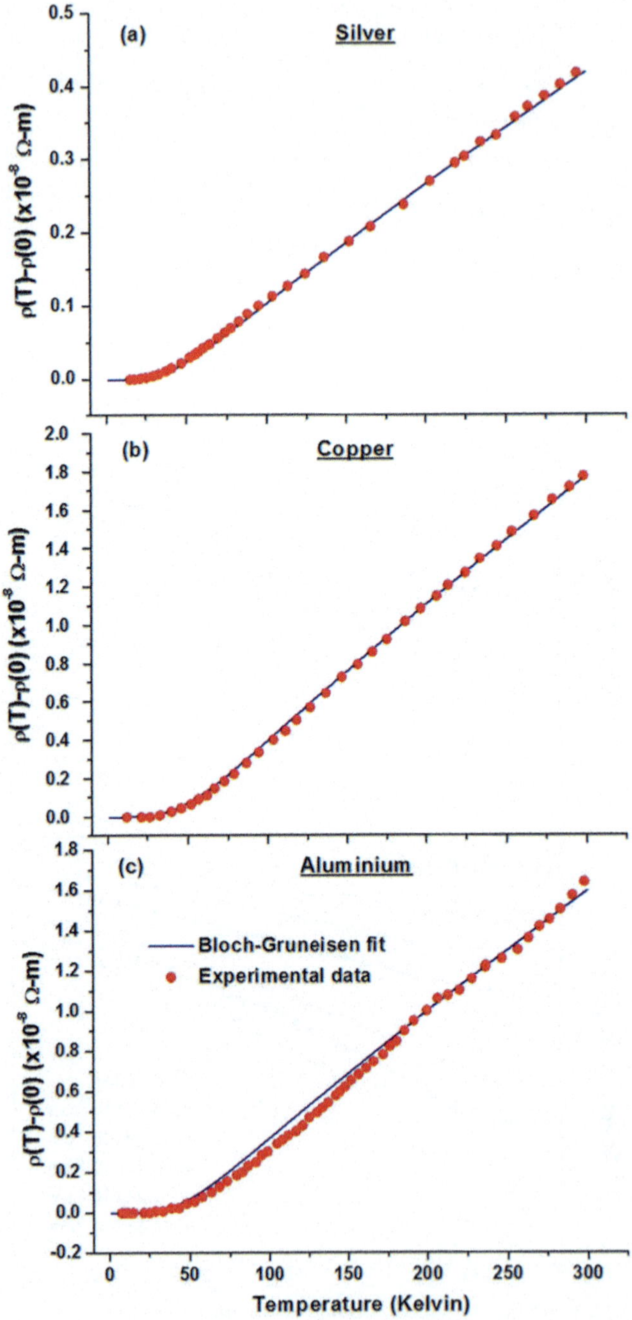

Fig. 8.2 Bloch-Grüneisen equation fitting of the experimental data for pristine films of **a** silver, **b** copper, and **c** aluminum thin films. Reprinted from Abhishek Chowdhury and Sudeep Bhattacharjee, J. Phys. D: Appl. Phys. 46 (2013) 435304), with permission of IOP publishing

Table 8.1 Summary of the results obtained in Fig. 8.1

Fluence (ion cm^{-2})	Silver		Copper		Aluminium	
	R_S at 15 K (Ω)	R_S at 295 K (Ω)	R_S at 15 K (Ω)	R_S at 295 K (Ω)	R_S at 15 K (Ω)	R_S at 295 K (Ω)
Pristine	0.014	0.298	0.112	1.887	0.309	2.107
8.7×10^{15}	0.191	0.475	0.27	2.046	0.519	2.317
9.69×10^{15}	0.453	0.868	0.459	2.234	0.93	2.728
1.12×10^{16}	0.801	1.633	0.871	2.647	1.401	3.199
1.24×10^{16}	1.306	2.24	1.59	3.366	1.854	3.652
1.38×10^{16}	1.689	2.873	2.265	4.04	2.376	4.174

in Fig. 8.1. On the other hand, for ion-irradiated films, the residual sheet resistance (R_{RS}), measured at 15 K, increases with the mass of projectile ions. The magnitude of R_s obtained by irradiating with various projectile mass and measured at two different temperatures is listed in Table 8.2.

8.3 Room Temperature to Residual Resistance (RRR) and Change in Residual Sheet Resistance (ΔR_S) for Ion Beam Irradiation with Varying Mass and Fluence

The residual resistivity ratio (RRR) for three metal samples (silver, copper, and aluminum) is depicted in Fig. 8.4a and b, plotted against ion beam fluence and mass number, respectively. As both fluence and mass number increase, the RRR exhibits a decreasing trend, with copper demonstrating the highest RRR among the three metals. Specifically, the RRR decline is less pronounced with varying fluence (from 7.6 to 1.6) as compared to the decrease with the mass number changes (from 12.4 to 1). For the copper sample, for instance, the RRR drops by 78.9% due to fluence changes, and by a more substantial 83.8% when mass is altered.

The changes in residual sheet resistances ($R_{RS} = (R_{RS})_{\text{irradiated}} - (R_{RS})_{\text{pristine}}$) for silver, copper, and aluminum are shown in Fig. 8.5a and b as a function of beam fluence and ion mass. The data has been fitted using a general empirical formula. The ΔR_{RS} values range from 0.05 to 2.2 Ω with increase in fluence for different metal thin films, and from 0.01 to 1.7 Ω with varying mass number. An exponential relation of the form $\Delta R_{RS} = C[e^{t/\alpha} - 1]$ describes well the variation of R_{RS} with fluence, where t represents the beam fluence (ions cm^{-2}). The values of the constants C and α used to fit the R_{RS}-fluence variation for different metals are provided in Fig. 8.6. Similarly, the variation of R_{RS} with the mass number of ionic species follows a parabolic trend, described by the equation: $\Delta R_{RS} = Am^2 + Bm$, where m represents

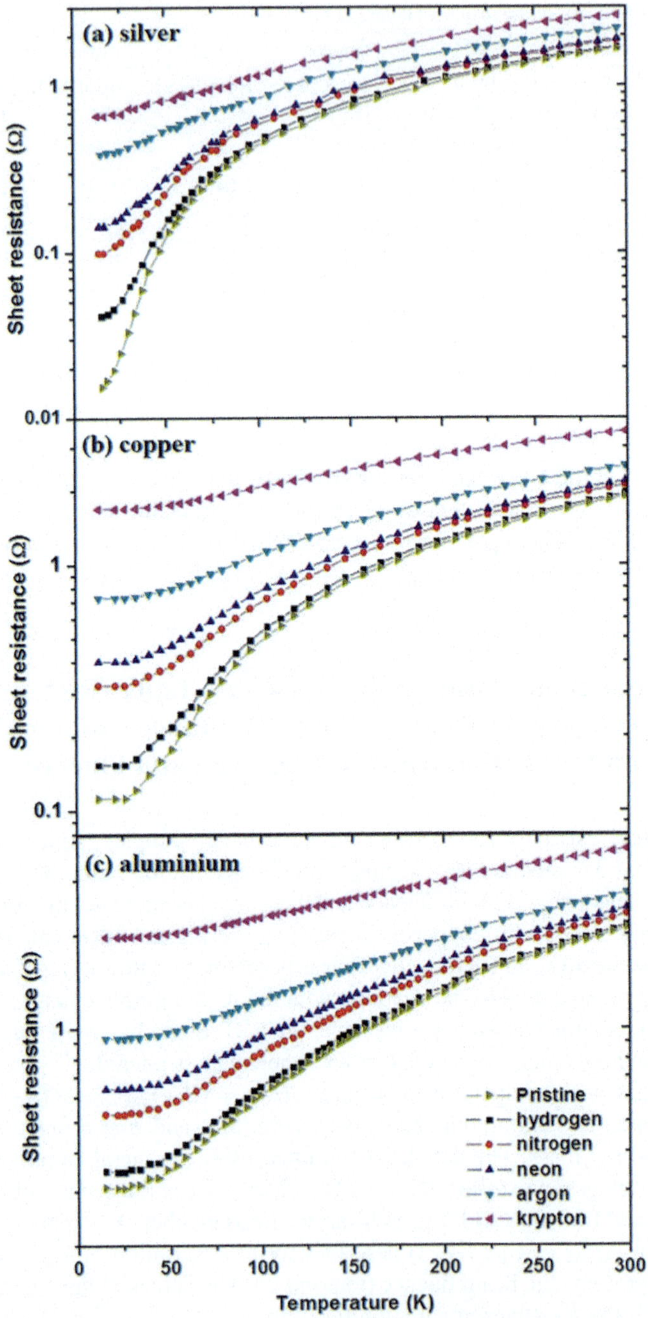

Fig. 8.3 Temperature variation of (R_S) of pure and irradiated films with varying mass number for **a** silver, **b** copper and **c** aluminum thin films. Reprinted from Abhishek Chowdhury and Sudeep Bhattacharjee, J. Phys. D: Appl. Phys. 46 (2013) 435304, with permission of IOP publishing

Table 8.2 Summary of the results obtained in Fig. 8.3

Mass No.	Silver		Copper		Aluminium	
	R_S at 15 K (Ω)	R_S at 295 K (Ω)	R_S at 15 K (Ω)	R_S at 295 K (Ω)	R_S at 15 K (Ω)	R_S at 295 K (Ω)
Pristine	0.014	0.298	0.112	1.987	0.309	2.117
H_2 (1)	0.191	0.475	0.27	2.032	0.519	2.323
N_2 (14)	0.453	0.868	0.459	2.123	0.93	2.718
Ne (20)	0.801	1.633	0.871	2.475	1.401	3.198
Ar (40)	1.306	2.24	1.59	3.386	1.854	3.602
Kr (83.8)	1.689	2.873	2.265	4.042	2.376	4.274

Reprinted from Abhishek Chowdhury and Sudeep Bhattacharjee, J. Phys. D: Appl. Phys. 46 (2013) 435304, with permission of IOP publishing

the atomic mass number of the ionic species in amu. The values of the constants *A* and *B* for different metals are listed in Fig. 8.6.

The fitting equations can be leveraged to predict the change in residual sheet resistance (R_{RS}) for a gaseous ion like argon at varying fluence levels, as well as for diverse mass numbers at a fixed fluence of 9.6×10^{15} ions cm^{-2}. Experimental measurements demonstrate a substantial change in sheet resistance (R_S) upon ion irradiation. For instance, when a copper film is subjected to krypton ion irradiation at room temperature, R_S increases from 1.987 to 4.042, constituting nearly a two-fold increase. Likewise, argon ion irradiation leads to an increase in R_S from 1.987 to 3.386, representing approximately a 1.7-fold rise.

8.4 Penetration Depth of the Ions Inside the Samples

The diffusion of implanted ions within the metallic thin film plays a pivotal role in determining electrical properties of the material and thus warrants a detailed investigation. To gain a deeper understanding of the variations in sheet resistance (R_S) induced by ion irradiation, it is essential to analyze the spatial distribution and penetration depth of the implanted ions within the copper film. Such analysis provides critical insights into the extent of ion diffusion, which directly influences the electrical behavior of the film. Cross-sectional energy-dispersive X-ray spectroscopy (EDX) was employed to investigate the distribution of implanted ions across the host material. EDX measurements were performed on copper film samples irradiated with argon ions, generating both line and map scans of the material. The line scan provides a detailed depth profile, illustrating the distribution of implanted Ar^+ ions and the structural profile of the copper film. The map scan, on the other hand, offers precise spatial localization of the Ar^+ ions within the film, enabling a comprehensive evaluation of ion diffusion and its impact on the film's resistive properties.

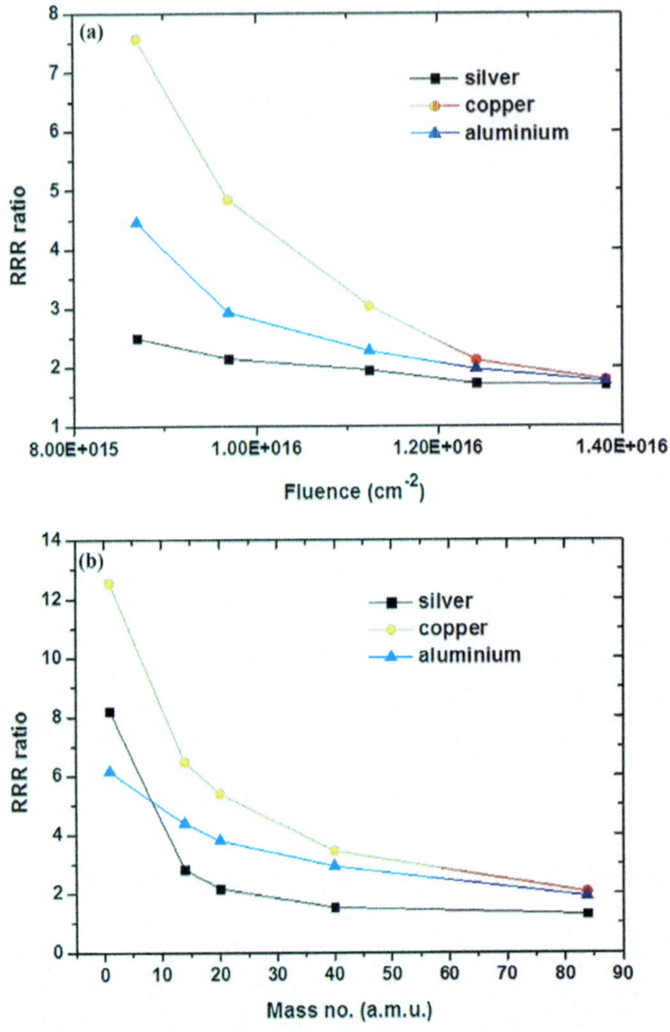

Fig. 8.4 Plot of RRR of the irradiated nano-films with **a** varying fluence and **b** varying ionic species. Reprinted from Abhishek Chowdhury and Sudeep Bhattacharjee, J. Phys. D: Appl. Phys. 46 (2013) 435304, with permission of IOP publishing

Figure 8.7a illustrates the line scan results for an argon-irradiated copper film deposited on a silicon substrate, subjected to an ion fluence of 1.24×10^{16} ions cm^{-2}. The scanned line, marked by the yellow line in the image, corresponds to a copper film with an approximate thickness of 175 nm. The experimental data reveals that Ar^{+} ions penetrate to a depth of about 75 nm within the copper film, significantly exceeding the penetration depth predicted by SRIM simulations, which estimate a depth of only ~ 2 nm. Figure 8.7b showcases EDX map scan data, which highlights

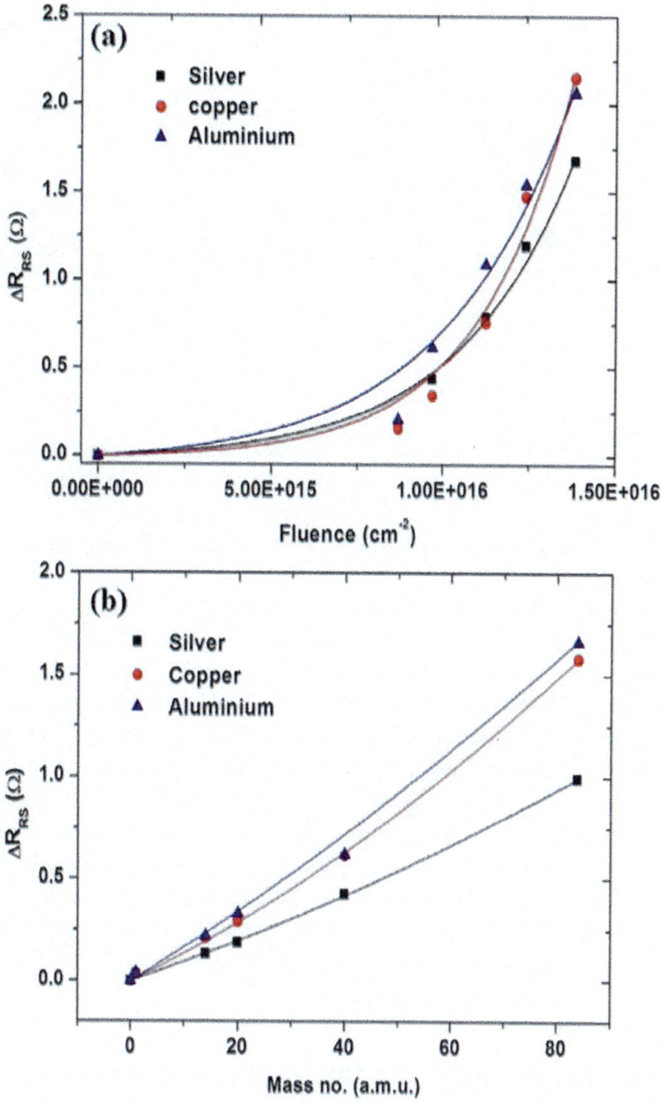

Fig. 8.5 Change in sheet resistance of irradiated silver, copper, and aluminum films with **a** varying fluence and **b** varying ionic species. Reprinted from Abhishek Chowdhury and Sudeep Bhattacharjee, J. Phys. D: Appl. Phys. 46 (2013) 435304, with permission of IOP publishing

the silicon K-α1, copper L-α1, and argon K-α1 lines in distinct colors, confirming the substantial penetration of Ar ions into the copper layer. This notable discrepancy between the experimentally observed penetration depth and SRIM simulation predictions underscores the need for a deeper investigation into the mechanisms driving ion diffusion. By utilizing the diffusion length obtained from EDX measurements, the

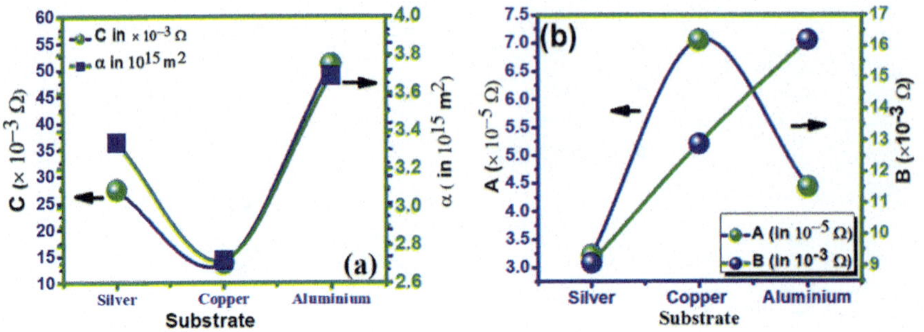

Fig. 8.6 Variation of fitting parameters for ΔR_{RS} with **a** fluence for different substrates. Fitting equation: $\Delta R_{RS} = C[\exp(t/\alpha) - 1]$, **b** mass number. Fitting equation: $\Delta R_{RS} = Am^2 + Bm$

diffusion coefficient of the Ar ions in the copper film can be calculated using Fick's law of diffusion. A thorough understanding of this diffusion process is essential for explaining the unexpectedly large penetration depths observed in the copper film, which could have significant implications for the modification of material properties under ion irradiation.

The diffusion length L_d is given by the formula $L_d = 2\sqrt{Dt}$, where D is the diffusion coefficient and t is the diffusion time, which is 2 seconds in our experiments, as the samples were irradiated for 2 s. For a diffusion length $L_d = 50$ nm, D is calculated to be 3.12×10^{-16} m² s⁻¹. Since reliable values for the diffusion coefficient of argon ions in copper substrates are scarce in the literature, these findings are compared with diffusion results from other experiments. For instance, at room temperature, the diffusion coefficient of hydrogen in nickel is approximately $\sim 2 \times 10^{-14}$ m^2 s^{-1} [1], and for gallium ions implanted in a silicon layer at 400 °C, the diffusion coefficient is 1.3×10^{-15} m^2 s^{-1} at 400 °C [2]. In another study, for low-energy (300 eV) Ar^+ irradiation in an (In, Ga) Sb heterojunction, the diffusion length L_d for ion bombardment-induced defects is around 100 nm [3]. These values are in a range comparable to those obtained in our experiments (Fig. 8.7).

8.5 Debye Temperature and the Elastic Modulus

The Bloch-Grüneisen formula indicates that the modified resistivity affects the Debye temperature (θ_D), which in turn influences the electron-phonon interactions in the sample. The temperature dependence of the experimental sheet resistance is fitted using Eq. (7.5) from the previous chapter (as shown in Fig. 8.2) to determine the Debye temperature of the sample. Figure 8.8a and b displays the variation of θ_D with ion beam fluence and ionic mass. The Debye temperatures of the pristine samples are close to the bulk values of the metals. However, it is observed that θ_D decreases with an increase in both fluence and mass number, although the nature of the decrease differs in each case. This reduction in Debye temperature is attributed to a weakening

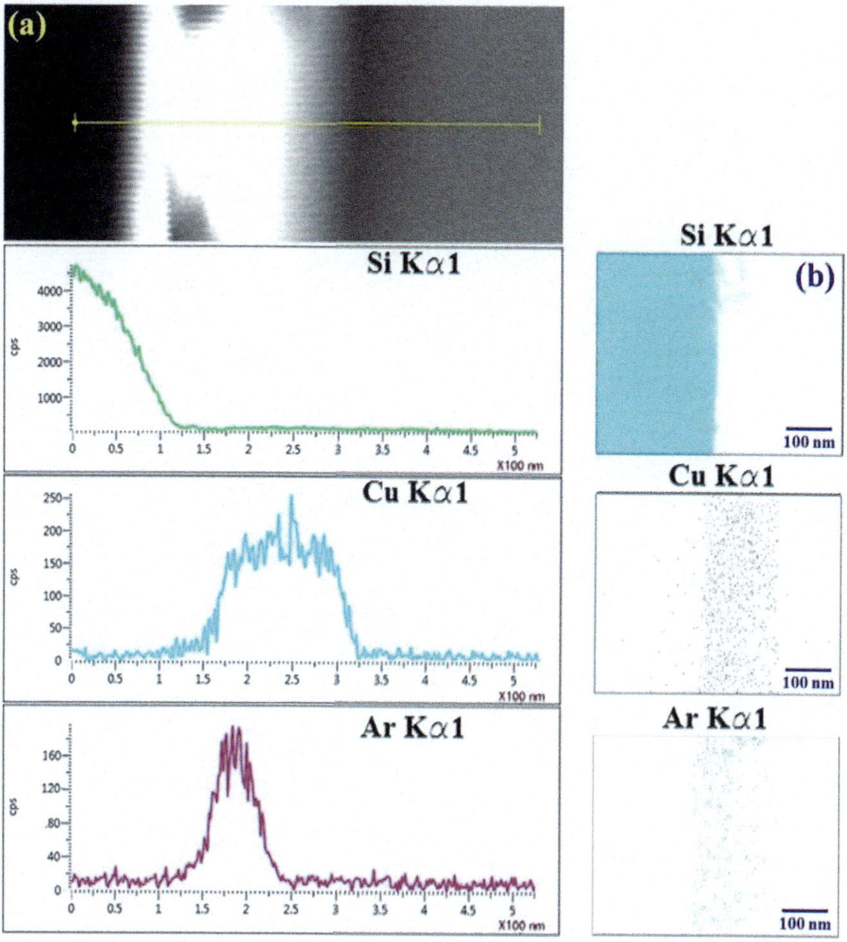

Fig. 8.7 **a** EDX line scan and **b** EDX map scan results for argon irradiated copper film on silicon substrate (beam fluence is 1.24×10^{16} ions cm^{-2}). Reprinted from Abhishek Chowdhury and Sudeep Bhattacharjee, J. Phys. D: Appl. Phys. 46 (2013) 435304, with permission of IOP publishing

of electron-phonon coupling, also referred to as phonon softening [4, 5], caused by the presence of implanted impurities.

Figure 8.9a and b illustrates the variation of the bulk modulus E with fluence and mass number, respectively, showing a trend similar to that of θ_D.

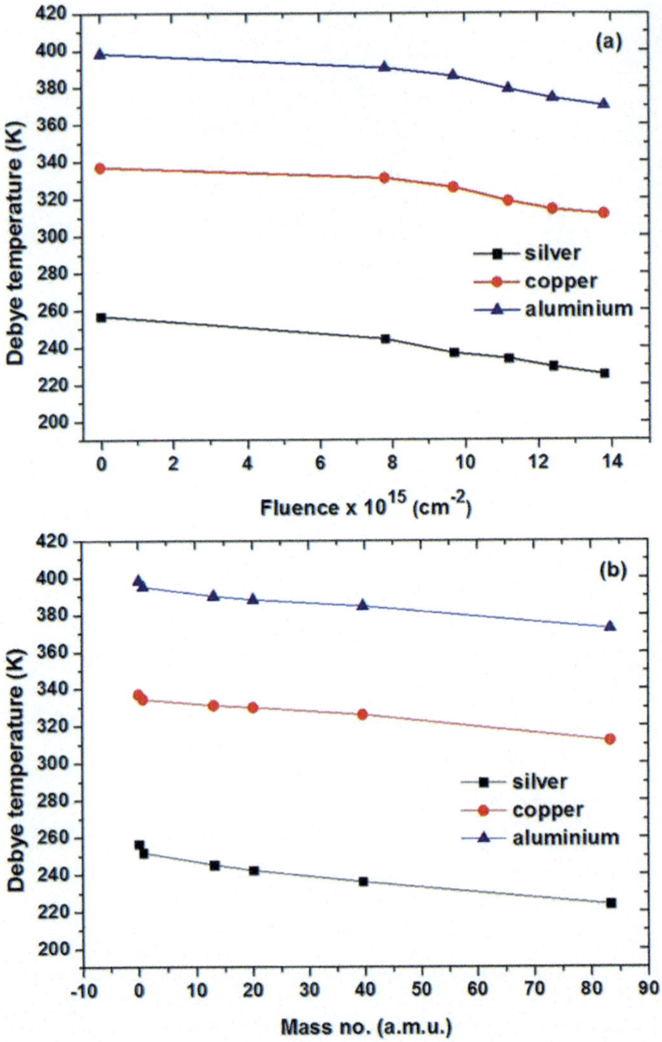

Fig. 8.8 Variation of Debye temperature for pure and irradiated samples as a function of **a** beam fluence and **b** ionic mass. Reprinted from Abhishek Chowdhury and Sudeep Bhattacharjee, J. Phys. D: Appl. Phys. 46 (2013) 435304, with permission of IOP publishing

8.6 A Discussion on the Diffusion of Ions in Matter and Available Simulation Tools

Due to the significantly shorter wavelength of energetic electron beams, energy-dispersive X-ray (EDX) mapping serves as both a quantitative and qualitative tool for detecting implanted projectile ions. Unlike the ion penetration depths predicted

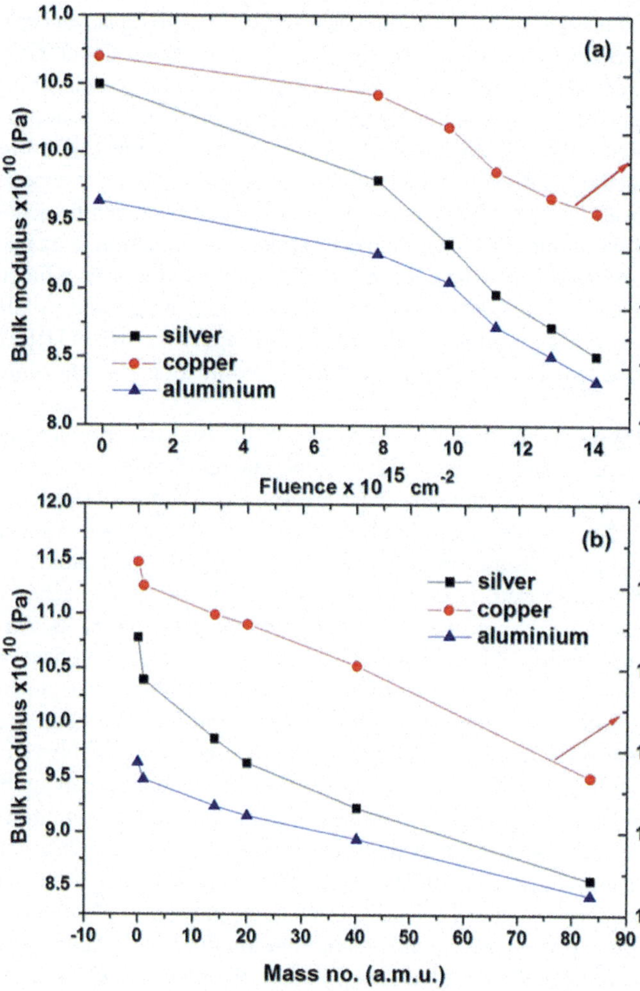

Fig. 8.9 Variation of bulk modulus for pure and irradiated samples as a function of **a** beam fluence and **b** ionic mass. Reprinted from Abhishek Chowdhury and Sudeep Bhattacharjee, J. Phys. D: Appl. Phys. 46 (2013) 435304, with permission of IOP publishing

by SRIM simulations, which are based on static Monte Carlo calculations, EDX mapping provides a more comprehensive assessment, as it allows for the detection of ions that have diffused much deeper into the substrate. This enhanced depth sensitivity makes EDX mapping particularly valuable for analyzing ion implantation beyond the theoretical range estimated by TRIM [6]. The Monte Carlo simulations performed using the SRIM (Stopping and Range of Ions in Matter)/TRIM (Transport of Ions in Matter) software offer a near-accurate estimation of ion penetration depth (implantation range) and the distribution of defects as a function of depth. Previous studies have demonstrated that the thickness of the damage layer (amorphous layer) predicted by SRIM simulations closely aligns with cross-sectional TEM observations. However, a key limitation of SRIM is that it does not account for the dynamic effects of ion irradiation, such as target layer sputtering and continuous implantation of projectile ions. To overcome these limitations, an alternative Monte Carlo simulation method, TRIDYN, which incorporates similar interaction potentials, has been introduced [7]. TRIDYN offers a more advanced approach by accounting for the dynamic evolution of the target material under ion irradiation. It enables the estimation of variations in atomic concentration of implanted ions as a function of ion fluence and provides insights into surface recession, i.e., the thickness of the material sputtered away during irradiation. This additional capability makes TRIDYN a more comprehensive tool for studying ion-induced modifications in thin films and bulk materials.

One of the primary reasons for the discrepancy between experimental findings and SRIM/TRIM simulation predictions lies in the inherent limitations of these computational tools, particularly at the specific energy levels and ion fluences used in the experiments. Both SRIM and TRIM are based on Monte Carlo simulations that primarily describe ion transport in an amorphous medium, where atoms are assumed to be randomly positioned. These models rely on a ballistic recoil mechanism (binary collision approximation), wherein individual ions undergo binary nuclear collisions with target atoms, displacing them and generating a cascade of defects. While this approach provides a reasonable approximation of ion penetration depth and damage distribution, it fails to capture several critical physical processes that occur in real crystalline materials, especially under low-energy irradiation and high ion fluences. A key limitation of SRIM/TRIM is that they do not account for many-body interactions, which play a crucial role in defect formation and evolution. Since these simulations treat ion-target interactions as a series of independent binary collisions, they neglect the cumulative effects of prior ion implantation, which can lead to local structural modifications, defect accumulation, and material evolution over time. This omission is particularly significant at high ion fluences, where the progressive buildup of defects and implanted species alters the response of target material to further irradiation. Furthermore, SRIM/TRIM simulations do not incorporate damage caused by inelastic processes, such as electronic excitations, ionization effects, and interactions with lattice phonons, which contribute to defect formation and relaxation mechanisms. These inelastic interactions can significantly impact the final defect distribution, particularly in metals and semiconductors where electron-phonon coupling is strong. Additionally, SRIM/TRIM do not account for thermally activated defect migration or ion beam-assisted relaxation effects, both of which influence the final microstructure of the irradiated material. As a result, these simulations provide only a rough approximation of defect formation, often underestimating the presence of both point defects (such as vacancies and interstitials) and extended defects (such as dislocation loops and stacking faults). The inability of SRIM/TRIM to model dynamic defect evolution, ion-induced mixing, and the long-range diffusion of implanted species further contributes to the observed mismatch between theoretical predictions and experimental measurements. Addressing these limitations requires either advanced simulations, such as molecular dynamics or hybrid Monte Carlo models, or direct experimental validation using techniques like cross-sectional TEM, EDX mapping, and Rutherford backscattering spectroscopy (RBS) to obtain a more accurate representation of ion-induced modifications in crystalline materials.

Another fundamental limitation of SRIM and TRIM simulations is their assumption that ions always penetrate a pristine, undisturbed target material. In reality, ion implantation is a dynamic process, and with increasing fluence, the target material undergoes significant structural modifications that these simulations fail to capture. Critical phenomena such as ion-induced surface erosion, the accumulation of lattice damage, and impurity agglomeration are not considered, leading to an oversimplified representation of the implantation process. Additionally, SRIM and TRIM do not account for local atomic rearrangements, including grain growth, phase transformations, or alloy formation, which can substantially alter the material's response to subsequent ion impacts. To overcome these limitations and improve the accuracy of ion penetration depth predictions at the specific energy levels and fluences used in our experiments, we employed the more advanced Monte Carlo-based ion implantation simulation software, TRIDYN HZDR [7, 8]. Unlike SRIM/TRIM, TRIDYN incorporates dynamic target evolution by considering ion-induced sputtering, implantation-induced material densification, and compositional changes over time. This allows for a more realistic modeling of high-fluence ion implantation, where progressive modifications to the target material significantly affect ion penetration behavior. Simulations performed under the experimental conditions described in Fig. 8.6 show that TRIDYN HZDR provides an improved estimation of ion penetration compared to SRIM/TRIM. While SRIM simulations predict a maximum penetration depth of approximately 2 nm for argon ions, TRIDYN HZDR increases this estimate to 4.5 nm, suggesting that dynamic target modifications indeed play a crucial role. However, despite this improvement, TRIDYN still falls short of explaining the experimentally observed deeper penetration of ~ 75 nm. This discrepancy indicates that additional physical mechanisms, such as radiation-enhanced diffusion, defect-mediated transport, or complex ion-material interactions, must be considered to fully understand the anomalous ion penetration depth in the copper film. Future studies integrating molecular dynamics (MD) simulations or hybrid Monte Carlo models may provide further insights into these complex ion-solid interactions. A potential explanation for the significantly greater penetration depth observed in experiments is the enhanced thermal diffusion of implanted ions, facilitated by the heat generated at the film surface during high-dose ion beam irradiation. Under intense irradiation, localized heating can elevate the temperature of the copper film, thereby promoting the deeper diffusion of implanted argon ions beyond the depths predicted by conventional simulations. This effect becomes particularly significant in materials with relatively high thermal conductivity, such as copper, where heat dissipation occurs rapidly but can still lead to localized temperature spikes at the nanoscale. The diffusion behavior of implanted species is governed by the temperature-dependent diffusion coefficient, $D(T)$, which follows an Arrhenius-type relationship: $D(T) = D_0\ e^{-(Q_d/RT)}$

where D_0 is the pre-exponential factor, Q_d represents the activation energy for diffusion, R is the universal gas constant, and T is the absolute temperature. This equation implies that even a moderate increase in temperature can significantly enhance atomic mobility. For instance, an increase of just 100 K above room temperature can amplify the diffusion coefficient by approximately three orders of magnitude, vastly increasing the depth to which implanted argon ions can migrate within the copper matrix. A similar phenomenon has been well-documented in boron implantation into silicon, where an initial rapid transient diffusion occurs in crystalline silicon samples before reaching equilibrium diffusion behavior [9]. Such transient diffusion effects could also contribute to the unexpectedly deep penetration of argon ions in the present study. Additionally, radiation-enhanced diffusion mechanisms, such as defect-assisted migration and ion beam-induced mixing, may further accelerate the redistribution of implanted ions. These effects, in combination with thermal diffusion, could explain the large discrepancy between experimental penetration depths and theoretical predictions based solely on ballistic collision models. Further investigations, such as in situ temperature measurements during ion irradiation and atomistic simulations incorporating thermal effects, would be essential for quantifying the precise contribution of thermal diffusion to the observed ion penetration behavior.

Questions

1. What are the ways in which one can vary the ion beam induced defects and impurities in a metallic thin film?
2. What is meant by residual sheet resistance of a metallic thin film?
3. What is the Bloch–Grüneisen equation for the variation of resistivity of a material with temperature?
4. Why is the resistivity of a thin metallic film much higher than the resistivity of the bulk material? Can you comment of the nature of the variation of the sheet resistance R_S of pristine films with temperature from room temperature (300 K) until about 70 K, and then below this temperature? Why is the behavior nonlinear below 70 K?
5. What does EDX diagnostics (energy dispersive X ray spectroscopy) tell us about a material? What does map and line scan of a sample imply? Why can EDX diagnostic predict elements present much deeper from the surface, compared to other conventional tools?
6. What is Fick's law of diffusion? Which physical quantity can be calculated employing Fick's law of diffusion? Calculate the diffusion length of Argon ions inside copper thin films for a diffusion time of 10 s and a diffusion coefficient of 3.12×10^{-16} m^2 s^{-1}.
7. What is phonon softening in electron-phonon interaction? How does the presence of impurities affect phonon softening?
8. What are the limitations of SRIM/TRIM simulations in properly addressing ion matter interactions?

9. How is the diffusion coefficient affected at higher temperatures? Calculate the diffusion coefficient for Boron implantation is Silicon at 300 and 1000 K. You may take Q_d the activation energy for diffusion as 3.21 eV.
10. What can you deduce from the different nature of variation of the residual sheet resistance $\Delta R_{RS} = (R_{RS})_{irradiated} - (R_{RS})_{pristine}$, with fluence and ionic mass, where the dependance have been found to be exponential and parabolic respectively.
11. A thin film of Indium Tin Oxide (ITO) with an initial sheet resistance of $R_s = 50\ \Omega/cm^2$ is exposed to an ion beam treatment using Ar ions at an energy of 100 keV and a fluence of 5×10^{15} ions/cm^2. After irradiation, the sheet resistance increases to 120 Ω/cm^2, then (a) what is the percentage change in the sheet resistance due to ion beam irradiation? (b) what possible physical mechanisms could explain this increase in sheet resistance? (c) how would further increasing the ion fluence likely affect the sheet resistance?

References

1. E. Wimmer, W. Wolf, J. Sticht, P. Saxe, C.B. Geller, R. Najafabadi, G.A. Young, Temperature-dependent diffusion coefficients from ab initio computations: hydrogen, deuterium, and tritium in nickel. Phys. Rev. B **77**, 134305 (2008)
2. R. Batabyal, S. Patra, A. Roy, S. Roy, L. Bischoff, B.N. Dev, Estimation of diffusion coefficient by photoemission electron microscopy in ion-implanted nanostructures. Appl. Surf. Sci. **256**, 536 (2009)
3. A.H. Eltoukhy, J.E. Greene, Diffusion enhancement due to low-energy ion bombardment during sputter etching and deposition. J. Appl. Phys. **51**, 4444 (1980)
4. G. Kastle, T. Muller, H.-G. Boyen, A. Klimmer, P. Ziemann, Resistivity and phonon softening in ion-irradiated epitaxial gold films. J. Appl. Phys. **96**, 7272 (2004)
5. G. Kastle, H.-G. Boyen, A. Schroder, A. Plettl, P. Ziemann, Size effect of the resistivity of thin epitaxial gold films. Phys. Rev. B **70**, 165414 (2004)
6. J.F. Ziegler, M.D. Ziegler, J.P. Biersack, SRIM-the stopping and range of ions in matter (2010). Nucl. Instrum. Methods Phys. Res. B **268**, 1818–1823 (2010)
7. W. Moller, W. Eckstein, J.P. Biersack, Tridyn-binary collision simulation of atomic collisions and dynamic composition changes in solids. Comput. Phys. Commun. **55**, 355 (1988)
8. www.hzdr.de/db/Cms?pOid=21578&pNid=0
9. T.O. Sedgwick, A.E. Michel, V.R. Deline, S.A. Cohen, J.B. Lasky, Transient boron diffusion in ion-implanted crystalline and amorphous silicon. J. Appl. Phys. **63**, 1452 (1988)

Chapter 9
Case Study 3: Effect of Low-Energy Ion Irradiation on Optical Properties of Metallic Films

An in-depth discussion on the use of low-energy plasma ion beams for atomic-scale material modification, with a focus on metals, dielectrics, and organic compounds are provided in the previous chapters. This approach has been shown to enable the creation of atomically heterogeneous systems, thereby facilitating precise control over optical, electrical, and surface characteristics. The ability to engineer such properties at the nanoscale has sparked considerable interest in the research community, given its potential implications for advanced material design and functional device applications [1–5]. When low-energy ion beams interact with a target material, the incident ions penetrate the subsurface layers, typically extending to a depth of a few tens of nanometers [2]. This bombardment induces modifications at the atomic level, leading to localized compositional and structural changes that significantly alter the overall properties of materials [2, 4, 5]. Such modifications are particularly relevant in the field of optics, where even slight variations in the atomic arrangement can lead to substantial changes in optical response. The optical behavior of the material is fundamentally described by its optical constants, specifically the real (n) and imaginary (k) components of the refractive index. These parameters are crucial in determining a wide range of optical and electronic properties, including light absorption, reflection, and transmission. Furthermore, they provide valuable insights into electronic band structure, polarizability, refractive index variations, and skin depth. The optical constants also influence other critical material characteristics such as optical conductivity, energy loss functions, and surface plasmon behavior, implying the factors that play a pivotal role in applications ranging from photonics to plasmonic sensing. Additionally, these parameters govern interband and intraband transitions, which are key to understanding and manipulating charge carrier dynamics in materials [6–15]. This chapter focuses on an important and emerging application of ion beam processing by modulating the optical properties of metallic thin films via low-energy ion irradiation. By carefully controlling the ion fluence, energy, and species, it is possible to achieve precise tuning of the optical response of the film, thereby enhancing or tailoring its reflectivity, absorption, and plasmonic

S. Bhattacharjee, *Atomically Heterogeneous Systems*,
https://doi.org/10.1007/978-981-95-9733-8_9

behavior. This capability holds significant promise for next-generation optical coatings, plasmonic devices, and other advanced photonic applications. The following sections will provide a detailed exploration of the underlying mechanisms, experimental methodologies, and potential technological implications of ion-beam-induced optical modifications in metallic thin films.

The optical behavior of metallic thin film (MTF) mirrors has garnered significant interest within the scientific community in recent years [16–19]. The unique optical properties of MTFs make them highly valuable for a range of applications, including perfect lenses, enhanced transmission through perforated metal structures [20], mirror coatings for space missions, photodetectors, optical and biosensors, data storage technologies, and photonic devices [21–24]. A key aspect of understanding the optical response of these materials lies in analyzing their frequency-dependent dielectric constants, which provide insight into phenomena such as lattice vibrations, free carrier absorption, interband transitions, and chemical bonding excitations [25, 26]. For metals like silver, copper, and gold, which possess a single valence electron, dielectric properties in the lower energy range of the electromagnetic spectrum are predominantly influenced by intraband electronic transitions within the conduction band. In this regime, the optical response is largely dictated by free electron behavior, offering critical information about electron scattering and mean free path. At higher energies, interband transitions occur, where electrons are excited from occupied d-bands to partially filled sp-bands, leading to characteristic variations in dielectric constants [10]. Fundamentally, key optical properties such as transmission, reflection, absorption, and dispersion are intrinsically linked to the dielectric response of the material. Understanding and controlling these parameters are essential for optimizing MTFs for advanced optical and photonic applications.

The incorporation of low-energy ions into the lattice of a host metal can lead to significant alterations in its optical properties [27, 28]. Zuccon et al. [29] investigated the impact of 4 keV helium ion bombardment on the optical characteristics of gold and iridium thin films using UV-VIS-NIR spectrophotometry. The helium ion flux in this study was comparable to the alpha particle exposure experienced by materials in solar winds over one, two, and four years. These findings have critical implications for spacecraft optical components, as they suggest strategies for designing materials that can withstand prolonged exposure to space environments without performance degradation. Similarly, research has been conducted on changes in transmissivity and absorptivity of metallic and semiconductor oxide thin films (HfO_2, Ta_2O_5, Nb_2O_5, TiO_2, SiO_2) subjected to implantation with 60 keV protons and gamma rays [30]. In another study, Wei et al. [31] examined the reflectivity of 500 nm aluminum films irradiated with protons at varying energies (60, 100, 130, and 160 keV). A notable reduction in reflectivity was observed across the 250–800 nm wavelength range at a critical fluence of 1×10^{17} cm^{-2}. Additionally, Kumar et al. [32] demonstrated that implanting 80 keV Ar^+ ions at a 30° grazing angle onto silver thin films on Si(100) substrates led to the formation of cone-like nanostructures at lower fluences, significantly influencing reflectivity. Axelevitch et al. [33] explored how different deposition techniques such as sputtering, thermal evaporation can affect the electrical and optical properties of metallic thin films (MTFs) composed of Au, Al, Ag, Cu, and Ti. Furthermore, Hooper and Sambles [34] provided theoretical insights into the influence of film thickness on the transmissivity of aluminum and silver films, corroborating experimental findings. While extensive studies have examined the effects of high-energy ion implantation on MTFs [31], a detailed and systematic analysis of the optical properties of atomically heterogeneous systems formed by low-energy ion beams remains lacking [35]. Given that optical components with metallic mirrors in space and nuclear energy applications are continuously exposed to both low- and high-energy particles, understanding their response to ion irradiation is crucial. A comprehensive investigation into their behavior under such conditions is essential to mitigate potential failures and ensure long-term operational stability.

This chapter presents a detailed investigation into the optical properties—reflectivity, transmissivity, and absorptivity—of metallic thin films (MTFs) composed of copper, silver, gold, and aluminum when subjected to low-energy (~ 0.5 keV) inert gaseous argon (Ar^+) ion irradiation. The study utilizes UV-VIS-NIR spectrophotometry to analyze the modifications induced by ion implantation. To further examine these changes, the optical constants, including the real and imaginary components of the refractive index, are determined for both pristine and irradiated films using the pseudo-Brewster angle technique. The interaction of low-energy Ar^+ ions with the metallic lattice leads to modifications in the dielectric function, which directly influence the refractive index and, consequently, the optical response of the MTFs. These alterations are interpreted using the Maxwell-Garnett effective medium theory, providing insight into the formation of atomically heterogeneous structures upon ion irradiation. Additionally, the Bruggeman effective medium approximation is employed to further analyze the optical behavior of the mixed media under ion beam exposure. The implications of these findings are discussed in the context of potential applications in advanced photonic and plasmonic devices.

Chapter 10 explores the impact of low-energy (~ 0.5 keV) inert gaseous argon (Ar^+) ion irradiation on the optical properties of metallic thin films (MTFs) composed of copper (Cu), silver (Ag), and aluminum (Al). The study focuses on key optical parameters, including the real (n) and imaginary (k) components of the refractive index, dielectric constants, skin depth, and optical conductivity. The optical constants of pristine MTFs are determined using the universal Kramers-Kronig (KK) analysis, with reflectivity measurements obtained via UV-VIS-NIR spectrophotometry serving as input data. These optical properties are evaluated across an energy range of ~ 1–4.96 eV (corresponding to wavelengths between 250 and 1200 nm). For ion-irradiated MTFs, the evolution of optical constants as a function of ion fluence is analyzed, including skin depth ($\delta = \lambda/2\pi k$), optical conductivity ($\sigma = nkc/\lambda$), and dielectric functions ($\epsilon_1 = n^2 - k^2$ and $\epsilon_2 = 2nk$). The Maxwell-Garnett (MG) effective medium approximation is applied to model the modified dielectric response of ion-irradiated films, characterizing the composite nature of the resulting system. Furthermore, the values of n and k derived from the MG approximation are cross-validated with those obtained using the pseudo-Brewster angle technique, previously discussed in an earlier chapter. Comparative analysis at four distinct laser wavelengths (405, 532, 632.8, and 670 nm) reveals a strong correlation between the two methods, reinforcing the reliability of the optical characterization.

The stability of aluminum (Al) and silver (Ag) metallic thin films (MTFs) was examined under helium ion irradiation, illustrated in Chap. 10. This study aims to simulate the impact of alpha particle exposure on spacecraft and satellite surfaces in space environments. Helium ions are implanted at varying fluences and energies ranging from 0.5 to 3 keV, allowing for an assessment of ion-induced modifications under different irradiation conditions. The selected fluence values correspond to expected exposure levels over a 4- to 6-year solar orbiter mission. To evaluate the optical degradation of MTFs, their reflectivity is measured across a broad spectral range, spanning from ultraviolet to near-infrared (200–2500 nm), both before and after helium ion implantation. This investigation provides insights into the potential of these MTFs for enhancing the durability and performance of optical components in space applications. Additionally, atomic force microscopy (AFM) is employed to analyze surface morphology and roughness variations induced by ion implantation. Computational simulations using the SRIM/TRIM framework offer further insights by predicting the distribution profile and penetration depth of helium ions within the Al and Ag MTFs. This comprehensive approach aids in understanding the long-term stability and functional reliability of metallic coatings in space environments.

9.1 Experimental Setup and Methodology

The experimental setup for developing ion beam from plasma and irradiating the samples are already provided in Chap. 3. However, for the sake of completeness, the image of the actual system is provided in Fig. 9.1.

The metallic thin films are stored in a vacuum desiccator after deposition, to prevent further oxidation and contamination from atmosphere caused due to air exposure. The preparation method for both pristine and Ar^+ ion-irradiated thin films is consistent across all metallic thin films (MTFs), with additional precautions taken for aluminum to minimize oxidation when exposed to air. Each sample has dimensions of approximately 1.25×1.25 cm^2. The prepared samples are secured on the collector inside the ion extraction chamber and subjected to a negative bias of 0.5 kV. This applied voltage facilitates the extraction of positively charged Ar^+ ions, which are then directed toward the samples, enabling penetration into the metallic medium to a depth of around 20 nm [2]. Ion irradiation is carried out with varying fluences, ranging from 1.6×10^{15} to 8.3×10^{15} cm^{-2}. To ensure uniform exposure across the entire sample surface, the extracted ion beam is maintained at a diameter of 2 cm.

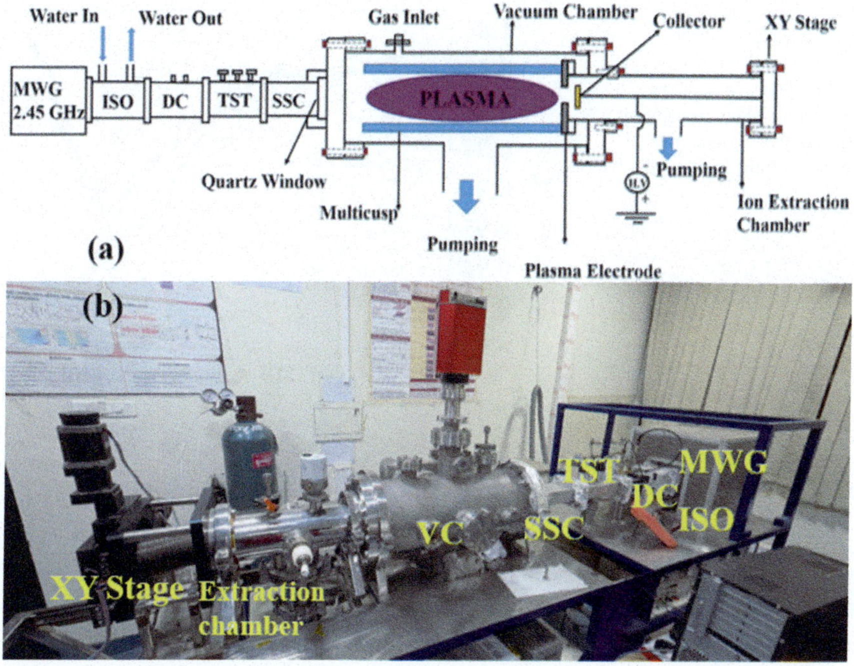

Fig. 9.1 **a** Schematic diagram and **b** digital picture of experimental setup. MWG, microwave generator; ISO, isolator; DC, directional coupler; TST, triple stub tuner; SSC, straight section; GI, gas inlet; and VC, vacuum chamber. Reprinted with permission from [5] © Optica Publishing Group

9.1.1 Measurement of Optical Constants

Figure 9.2 illustrates the variation of reflectivity (R_P) with the incidence angle of light (θ_i) of un-bombarded MTF (as grown) measured using the pseudo-Brewster angle method. The experimentally obtained data as well as fitted data (smoothened) of R_P versus θ_i are shown in Fig. 9.2. The obtained pseudo-Brewster angle (θ_B) and corresponding reflectivity of p-polarized light (R_{PB}) from the experiment (Fig. 9.2) are used to estimate the optical constants of MTF using Fresnel's equations as discussed later in Chap. 10.

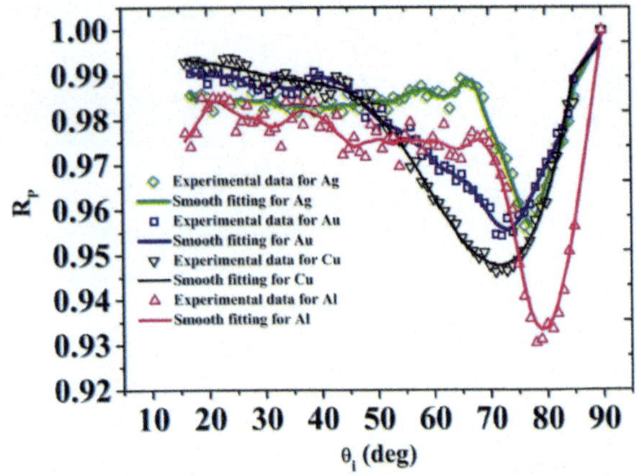

Fig. 9.2 Reflectivity of p-polarized light with varying incidence angles for silver, gold, copper, and aluminum MTF. Reprinted with permission from [5] © Optica Publishing Group

9.2 Optical Properties of Pristine and Ion Irradiated Metallic Thin Films (MTFs) Using 0.5 keV Ar^+ Ion Beams

9.2.1 Refractive Index (Real and Imaginary Parts) of Pristine MTFs Using the Pseudo-Brewster Angle Technique

The optical constants of pristine and ion-modified MTFs are estimated using pseudo-Brewster angle technique, described in Chap. 5. The real and imaginary components of the refractive index for both pristine and Ar^+ ion-irradiated metallic thin films (MTFs) of copper, silver, aluminum, and gold are presented in Fig. 9.3. In each plot, the left axis represents the real part, while the right axis corresponds to the imaginary part. The magnitudes of the refractive index for pure metals reported in previous literatures [7, 9, 36–40] show slight variations due to differences in measurement techniques and method of sample preparation, as noted in Ref. [33]. Even minor discrepancies in fabrication processes can lead to variations in the measured optical constants. In present case, the "zero fluence" corresponds to pristine (un-irradiated) MTFs, and the optical constants of pristine MTF exhibit strong agreement with the same reported by Johnson and Christy [39] and other experimental investigations [7, 9, 36–38, 40]. The magnitude of optical constants of pristine MTF in the present case are reasonably good when compared with existing literatures (the deviation in the imaginary part is 1.4–3% and the real is 15–20%). This validates the reliability of this experimental approach for investigation of irradiated samples. Experimental findings indicate that both the real and imaginary parts of the refractive index decrease progressively as the Ar^+ ion beam fluence increases. This declining trend aligns well with predictions from effective medium models [41, 42] and the Maxwell-Garnett (M-G) theory for mixed media [43]. For instance, in the case of copper MTFs, the real part of the refractive index decreases from 0.103 in pristine samples to 0.079 at

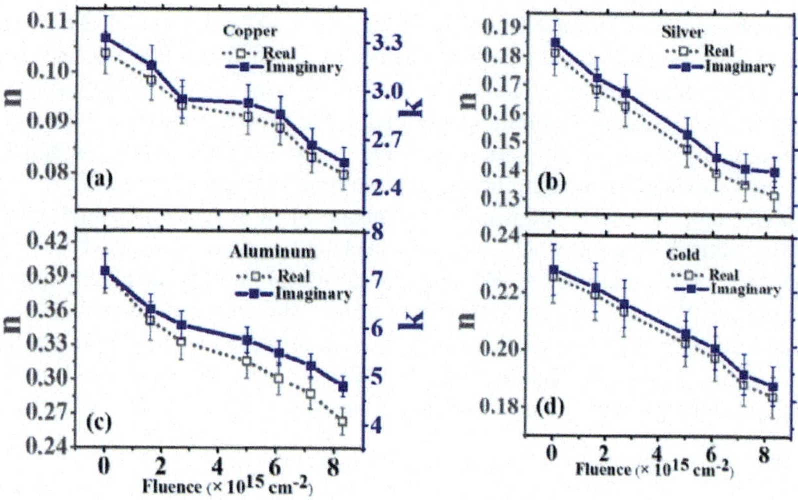

Fig. 9.3 Variation of the refractive index at 632.8 nm for **a** copper, **b** silver, **c** aluminum, and **d** gold with a varying fluences of argon ions as shown in blue for imaginary part on the right axis and black for real part on the left axis. Reprinted with permission from [5] © Optica Publishing Group

the highest fluence of 8.3×10^{15} cm^{-2}. A similar reduction is observed in case of imaginary part of the refractive index at maximum fluence. Silver and gold MTFs exhibit an almost linear decline in optical constants with increase in ion fluence, while aluminum demonstrates the most significant variation in the imaginary part of the refractive index under irradiation.

9.2.2 Reflectivity, Transmissivity, and Absorptivity of Ion Irradiated Metallic Thin Films

1. *Reflectivity, Transmissivity, and Absorptivity of Copper MTF*

The optical characteristics of copper metallic thin films (MTFs) under varying Ar^+ ion fluences are illustrated in Fig. 9.4. The samples were subjected to ion fluences of (1.6, 5.0, and 7.2) $\times 10^{15}$ cm^{-2}. A key observation from the experimental data in Fig. 9.4a (reflectivity) and Fig. 9.4b (transmissivity) is that these optical properties can be effectively tuned by controlling the concentration of implanted Ar^+ ions. As the ion fluence increases, the reflectivity of copper MTFs decreases. Notably, a sharp transition occurs around 544 nm, beyond which the reflectivity stabilizes. Theoretical studies indicate that interband transitions significantly influence the dielectric properties of metals and help explain the observed optical behavior. According to Segall's band structure calculations for copper [44], interband transitions occur near specific symmetry points in the Brillouin zone at approximately 2.25 eV and

4.1 eV, corresponding to wavelengths of 551 nm and 302 nm, respectively. These interband transitions alter the effective electron mass relative to the free electron mass, thereby increasing the effective electron density (n_{eff}). This change results in the characteristic reflectance minima observed in the experimental data, validating the theoretical predictions [45]. As seen in Fig. 9.4b, the transmissivity of copper MTFs increases with rising ion fluence, exhibiting a pronounced surge at 570 nm. This wavelength-dependent transmissivity enhancement suggests potential applications in optical filtering, where specific wavelengths could be selectively transmitted while others are blocked, enabling the design of narrow bandwidth optical filters. Figure 9.4c presents the absorptivity trends, showing that absorption rises with increasing Ar^+ ion fluence. This indicates that as more ions penetrate into the metallic medium, the material becomes increasingly absorptive and transmissive while experiencing a reduction in reflectivity.

2. *Reflectivity, Transmissivity, and Absorptivity of Silver MTF*

Silver exhibits higher electrical conductivity compared to copper, which is clearly reflected in its optical response. The reflectivity and transmissivity spectra of silver thin films are depicted in Fig. 9.5a and b, respectively. A notable feature in the reflectivity spectrum, shown in Fig. 9.5a, is the sharp increase in reflectivity beyond 316 nm. This abrupt transition can be attributed to interband transitions, as reported in previous studies [7, 9, 37–39]. Theoretical calculations indicate that interband transitions in silver occur around 3.8 eV [45], which corresponds to a wavelength of approximately 326 nm. This magnitude aligns closely with the experimentally observed transition at 316 nm. As the fluence of implanted Ar^+ ions increases, the overall concentration of embedded ions within the metallic thin film rises, leading to a reduction in the dielectric constant (or refractive index) of the modified film, as predicted by the Maxwell-Garnett (M-G) theory [43]. This theoretical framework provides a means to determine the effective dielectric properties of a heterogeneous medium, and its predictions are consistent with the experimentally obtained refractive index values (Fig. 9.3). The reflectivity of silver thin films decreases progressively with increasing ion fluence, as evidenced by the reflectance curve. However, the reduction in reflectivity after ion implantation is comparatively smaller for silver than for copper, with a decrease of approximately 20% from the maximum reflectivity of the pristine film. In the case of transmission, pristine silver films exhibit a distinct peak near 328 nm. Upon ion implantation, the transmissivity increases with ion fluence, and the peak shifts slightly toward longer wavelengths as more ions become embedded within the metallic matrix. Interestingly, this peak shift is unique to silver and is not observed in any of the other metallic thin films studied. Additionally, two minor peaks appear in the transmission spectra of the irradiated samples, becoming more pronounced at higher fluences. These peaks are absent in the pristine sample. The transmissivity initially increases with fluence, reaches a maximum value, and then undergoes a rapid decline before stabilizing. Absorptivity trends, illustrated in Fig. 9.5c, reveal a sharp reduction in absorption at approximately 312 nm. Notably, the cutoff wavelengths observed in the reflectivity, transmissivity, and absorptivity

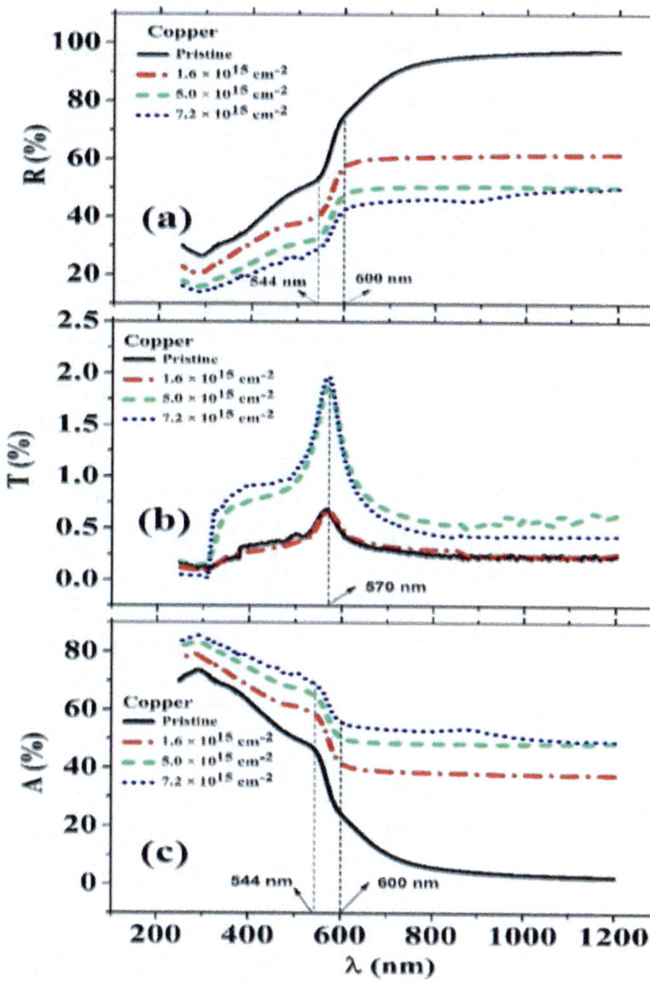

Fig. 9.4 **a** Reflectivity (*R*), **b** transmissivity (*T*), and **c** absorptivity (*A*) measured at 8° of normal incidence of copper metallic thin films irradiated with varying fluences of low-energy argon ions. Reprinted with permission from [5] © Optica Publishing Group

spectra are closely aligned, further confirming the consistency of the experimental results.

3. *Reflectivity, Transmissivity, and Absorptivity of Gold MTF*

Figure 9.6a depicts the decreasing magnitude of reflectivity of gold during Ar-ion bombardment with increasing fluence. A notable rise in reflectivity occurs around 480 nm. Gold exhibits at least two prominent interband transitions, occurring at approximately 330 nm and 470 nm in the near-UV and visible regions, respectively. These transitions impose limitations on the classical Drude model, which primarily

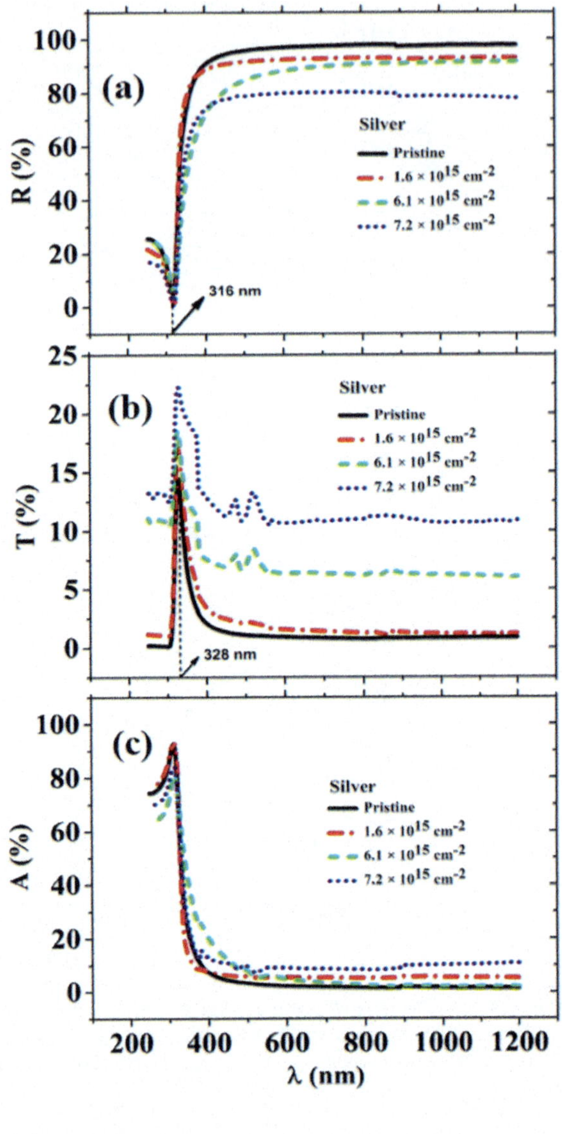

Fig. 9.5 **a** Reflectivity (R), **b** transmissivity (T), and **c** absorptivity (A) measured at 8° of normal incidence of silver metallic thin films irradiated with varying fluences of low-energy argon ions. Reprinted with permission from [5] © Optica Publishing Group

accounts for intraband transitions [9]. The interband transition wavelengths align well with the reflectivity minima observed for gold in Fig. 9.6a. Interestingly, the positions of these transitions remain unchanged across all ion-irradiated samples, indicating that fluence does not influence their spectral location. However, the magnitude of reflectivity reduction is quite significant, reaching approximately 30% at the highest fluence of 8.3×10^{15} cm^{-2}. The transmissivity spectrum of gold, depicted in Fig. 9.6b, reveals two broad peaks centered around 474 and 518 nm. These peaks persist across both pristine and irradiated samples, exhibiting no shift in their spectral positions. However, with increasing fluence, the overall transmission increases. Following the second peak, transmissivity reaches a saturation point. Absorptivity trends, as presented in Fig. 9.6c, demonstrate a characteristic increase with rising Ar^+ ion fluence. This suggests that as more ions penetrate the metallic film, the material becomes increasingly absorptive, a behavior consistent with modifications in electronic structure induced by ion implantation.

4. *Reflectivity, Transmissivity, and Absorptivity of Aluminum MTF*

Copper (Cu), silver (Ag), and gold (Au) are noble metals characterized by filled d-orbitals and valence electrons in the s-orbital. In contrast, aluminum (Al) has a distinct electronic configuration, with its valence electron occupying the p-orbital of the third shell while the s-orbital of the same shell remains fully occupied. In the reflectivity spectrum of aluminum presented in Fig. 9.7a, a noticeable dip is observed around 814 nm, corresponding to an energy of 1.52 eV. This aligns precisely with a known interband transition in aluminum [46]. The position of this dip remains unchanged despite increasing Ar^+ ion fluence, indicating that the implantation does not shift the interband transition energy. With increasing fluence levels of (1.6, 2.7, 7.2) $\times$ 10^{15} cm^{-2}, reflectivity consistently decreases. However, the decline is more pronounced in the ultraviolet (UV) and visible (VIS) regions compared to the near-infrared (NIR) range. As shown in Fig. 9.7b, transmissivity generally decreases with increasing Ar^+ ion fluence, except at 2.7×10^{15} cm^{-2}, where an unexpected rise is observed. This behavior is distinct from that of noble metals. Across all samples, both pristine and irradiated, transmissivity remains within $\sim$0–0.65%, indicating that aluminum thin films exhibit minimal transmission, with no sharp peaks appearing in the spectra. The impact of ion irradiation on transmissivity is therefore negligible. Absorptivity trends, illustrated in Fig. 9.7c, reveal a peak around 814 nm, mirroring the dip seen in the reflectivity spectrum. For both pristine and irradiated samples, absorptivity initially increases at this wavelength and subsequently declines. Overall, absorptivity rises with increasing Ar^+ ion fluence, suggesting that ion implantation enhances energy absorption within the material.

9.2.3 Optical Properties of Irradiated MTF Investigated through Mixed Medium Models—Maxwell-Garnett and the Bruggeman Approximations

This section presents a theoretical framework that qualitatively explains the experimental observations by: (1) accounting for impurity-induced modifications to the dielectric constant through the Maxwell-Garnett (M-G) and Bruggeman models, (2) elucidating the relationship between optical constants and reflectivity, and (3) exploring the effects of skin depth and transmissivity. To determine the effective

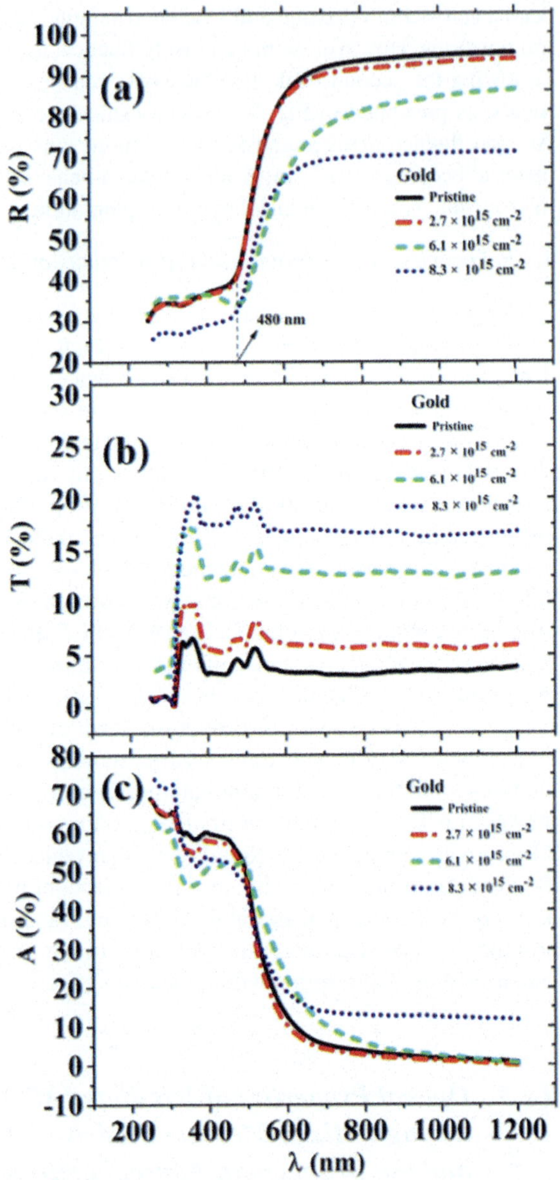

Fig. 9.6 **a** Reflectivity (R), **b** transmissivity (T), and **c** absorptivity (A) measured at 8° of normal incidence of gold metallic thin films irradiated with varying fluences of low-energy argon ions. Reprinted with permission from [5] © Optica Publishing Group

dielectric constant of a heterogeneous medium, several approximation techniques—such as M-G, Bruggeman, and the Extended Effective Medium Approximation (EEMA)—are commonly used. However, these models have inherent limitations based on inclusion size. As the inclusion dimension increases, their accuracy diminishes, as discussed in Ref. [47]. Reference [48] further demonstrates that when the inclusion radius is $r = 0.215\lambda$, where λ represents the incident radiation wavelength,

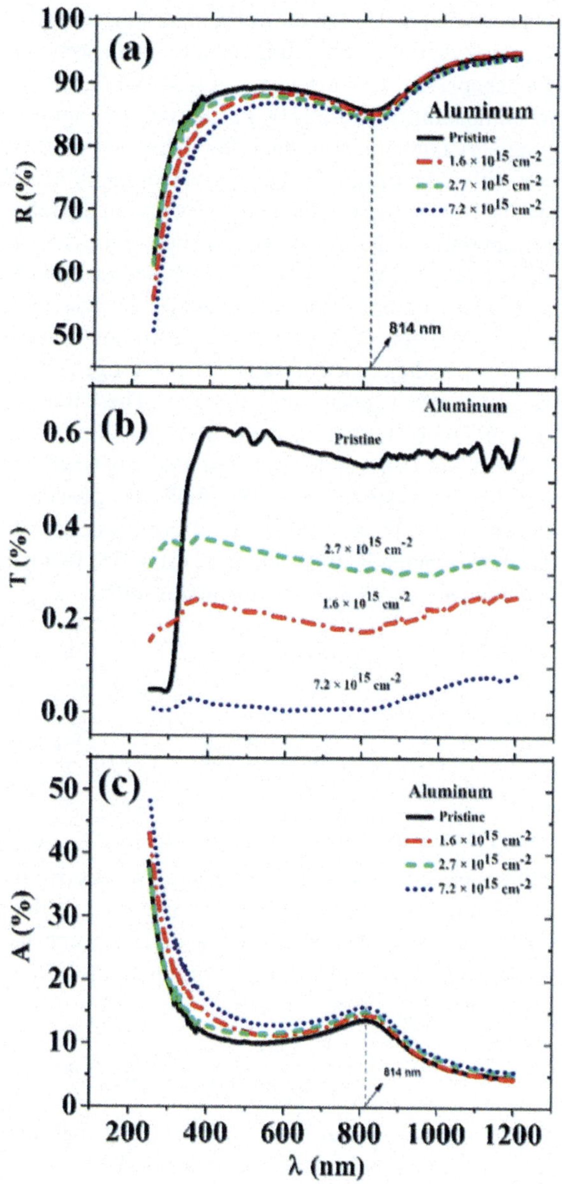

Fig. 9.7 **a** Reflectivity (R), **b** transmissivity (T), and **c** absorptivity (A) measured at 8° of normal incidence of aluminum metallic thin films irradiated with varying fluences of low-energy argon ions. Reprinted with permission from [5] © Optica Publishing Group

both M-G and Bruggeman approximations align well with Maxwell's equation-based models. However, for larger inclusions $r = 0.464\lambda$, these approximations diverge from the model, whereas EEMA provides a more accurate representation. In this study, the Van der Waals radius of argon atoms ($r = 0.188$ nm) is significantly smaller than the incident light wavelength (250–1200 nm), justifying the use of the M-G approximation. While previous studies have primarily focused on metallic inclusions embedded in dielectric matrices, the M-G model can also be applied to dielectric particles containing metallic inclusions, subject to specific constraints on size, shape, and loss mechanisms [42]. When the loss characteristics of inclusions are comparable to those of bulk materials, this approximation remains valid, with the imaginary component of the refractive index effectively describing energy dissipation within the medium. The presence of metallic inclusions in a dielectric medium results in an increase in the imaginary part of the refractive index, thereby enhancing absorption. Simultaneously, the real part of the refractive index can exceed that of the individual components. Our results, based on M-G approximation calculations, are consistent with prior studies on metal-dielectric composites [47, 49]. The observed variations in optical properties due to Ar^+ ion irradiation can be interpreted through the effective dielectric constant of the heterogeneous medium, estimated using either the M-G or Bruggeman approximation. These models define the microstructure of a random unit cell differently: the M-G model represents it as a coated sphere with a dielectric core surrounded by a host medium, whereas the Bruggeman model assumes a probability distribution for the dielectric constant, where the inclusion occupies a fraction f, and the host medium occupies: $1 - f$ [41]. The effective dielectric constant of heterogeneous media is estimated using the following formulas [41, 43] and compared with the experimental results:

$$\tilde{\epsilon}_{\mathrm{eff}}^{\mathrm{MG}} = \tilde{\epsilon}_{\mathrm{m}} + 3f\tilde{\epsilon}_{\mathrm{m}}\frac{(\epsilon_{\mathrm{Ar}} - \tilde{\epsilon}_{\mathrm{m}})}{\epsilon_{\mathrm{Ar}} + 2\tilde{\epsilon}_{\mathrm{m}} - f(\epsilon_{\mathrm{Ar}} - \tilde{\epsilon}_{\mathrm{m}})}, \tag{9.1}$$

$$f\frac{\left(\epsilon_{\mathrm{Ar}} - \tilde{\epsilon}_{\mathrm{eff}}^{\mathrm{Br}}\right)}{\epsilon_{\mathrm{Ar}} + 2\tilde{\epsilon}_{\mathrm{eff}}^{\mathrm{Br}}} + (1-f)\frac{\left(\epsilon_{\mathrm{m}} - \tilde{\epsilon}_{\mathrm{eff}}^{\mathrm{Br}}\right)}{\tilde{\epsilon}_{\mathrm{m}} + 2\tilde{\epsilon}_{\mathrm{eff}}^{\mathrm{Br}}} = 0, \tag{9.2}$$

where $\tilde{\epsilon}_{\mathrm{eff}}^{\mathrm{MG}}$ is the complex effective dielectric constant of the mixed medium using M-G approximation, $\tilde{\epsilon}_{\mathrm{eff}}^{\mathrm{Br}}$ is the complex effective dielectric constant of the mixed medium using Bruggeman approximation, $\tilde{\epsilon}_{\mathrm{m}}$ is the complex dielectric constant of the host metallic medium, ϵ_{Ar} is the dielectric constant of the inclusive argon atoms, and f is the fractional volume of argon atoms. The complex refractive index of the mixed medium is then given by

$$\tilde{n}_{\mathrm{eff}} = \sqrt{\tilde{\epsilon}_{\mathrm{eff}}} = n + ik, \tag{9.3}$$

where $\tilde{n}_{\mathrm{eff}}$ and $\tilde{\epsilon}_{\mathrm{eff}}$ are both complex. The real and imaginary parts of the effective refractive index are calculated using both M-G and Bruggeman approximation methods from Eqs. (9.1)–(9.3) and compared with the experimental results. It has been observed from the comparison results of copper, gold, silver, and aluminum that for the lower Ar^+ ion fluence, both M-G and Bruggeman approximations slightly deviate from the experimental values, while this deviation from the experimental values becomes larger as the fluence of Ar^+ ion beam increases. It is found that the deviation of numerical values of the refractive index is slightly higher in the case of the Bruggeman approximation compared to the M-G approximation for higher fluence. Hence, M-G approximation is found closer to the experiment. The number density and fractional volume f ($f = nv$, n = number density of Ar^+ ions, v = volume of each Ar^+ ion) of Ar^+ ions increase with Ar^+ ion fluence. As f increases, it results in an increase and decrease in real and imaginary parts of the $\tilde{\epsilon}_{eff}$, respectively, and therefore, both n and k decrease. The effect of voids and surface roughness on real and imaginary parts of dielectric constants of gold and bismuth, respectively, using the approximation methods is discussed in Refs. [50, 51]. It is shown in Ref. [50] that as fractional volume or volume ratio of inclusion (surface roughness and voids) increases inside gold, the real part of the dielectric increases, while the imaginary part of the dielectric decreases.

9.3 Discussion on the Results

The non-monotonic variations in reflectivity, transmissivity, and absorptivity of silver, gold, and aluminum under Ar^+ ion irradiation can be attributed to changes in root-mean-square (RMS) surface roughness and grain size induced by ion bombardment. While an increase in fluence generally leads to a rise in surface roughness, studies by Zuccon et al. [29] have shown that for specific He^+ ion fluences, the RMS roughness of gold films exhibits a slight reduction. In fact, the roughness of the most heavily irradiated gold samples was found to be lower than that of pristine films, accompanied by an increase in grain size. This grain growth is primarily due to energy deposition on the surface, which enhances adatom mobility and promotes coalescence. The anomalous behavior observed in gold reflectivity aligns with these morphological changes. A similar trend was observed in the atomic force microscopy (AFM) images of silver films, where surface roughness exhibited an unexpected decrease at higher fluences compared to lower fluences. This suggests that structural modifications in silver, gold, and aluminum play a critical role in the observed non-monotonic optical responses. Such effects are particularly pronounced in silver, aluminum, and gold, where changes in morphology strongly influence their reflectivity, transmissivity, and absorptivity. The theoretical framework outlined earlier provides a qualitative explanation for these experimental observations by: (1) incorporating the impact of impurities on the dielectric constant through the Maxwell-Garnett (M-G) and Bruggeman approximations, (2) elucidating the correlation between optical constants and reflectivity, and (3) analyzing the role of skin depth and transmissivity in modifying optical behavior.

For simplicity, the reflectivity of metallic media for normal incidence was tested to verify the effect of Ar^+ ion fluence. It is defined by the following relation:

$$R = \frac{(n_o - n)^2 + k^2}{(n_o + n)^2 + k^2}, \tag{9.4}$$

where n_o is the refractive index of incident media. The lowering in refractive indices n and k with Ar^+ ion fluence causes the reflectivity of metallic media to decrease, which can be verified from Eq. (9.4) and further confirmed by the experimental results of reflectivity presented for copper, silver, aluminum, and gold. Skin depth (δ) of an electromagnetic wave inside the absorbing medium is defined as the distance over which the amplitude of the electric field reduces to $1/e$ of its value on the surface, and it is given by $\delta = \lambda/2\pi k$, where λ is the wavelength of the incident radiation in vacuum, and k is the imaginary part of the refractive index of metallic media. It is found that k decreases with ion fluence, which in turn enhances the skin depth inside the metallic medium. The enhancement of skin depth with ion fluence would imply an increase in transmission of light through the medium.

Questions

1. In what ways can low energy ion irradiation change the optical properties of metallic thin films?
2. What experimental technique is employed to measure the real and imaginary parts of the refractive index?
3. What is the Kramers-Kronig (KK) technique? How can optical constants of pristine metallic thin films be obtained by employing the KK technique?
4. What optical properties of a material can be determined by employing UV-VIS-NIR spectrophotometry? What are the typical range of wavelengths that can be employed in this diagnostic?
5. What are the main theoretical models employed in the research to investigate the optical properties of mixed media? Can you describe the limitation of the models and how well they explain the experimentally obtained optical properties?
6. Ions arrive at the plasma sheath edge with Bohm velocity. Find out on what parameters does the velocity depend upon.
7. What is Pseudo-Brewster's angle? What does the minimum point in the experimental data of R_P versus θ_i (shown in Fig. 9.2) indicate? Why does both the real and imaginary parts of the refractive index decrease with increase in ion beam fluence?
8. What are the interband transitions in a metal? What role do they play in determining the optical (dielectric) constants of a metal? Where do the interband transitions lie for copper? What are their energies (or wavelengths)?
9. Where does the interband transition for silver occur, mention the energy and the corresponding wavelength?
10. What interesting feature is observed in the transmission peak of Ag with increase in ion beam fluence, that is not observed in any of the MTFs of other metals?
11. What are the two interband transitions in Gold?
12. What is the main characteristic difference in the electronic structure of Cu, Ag, and Au as compared to Al? Where does the interband transmission for Al occur?
13. What are the considerations for the applicability of the mixed medium models such as Maxwell-Garnet and Bruggeman to describe optics of mixed media? Comment on the agreement of the data of the dielectric constants obtained

experimentally for Cu, Ag, Au and Al, with those predicted by the two models, as a function of ion beam fluence.

References

1. I.P. Jain, G. Agarwal, Ion beam induced surface and Interface engineering. Surf. Sci. Rep. **66**, 77–172 (2011)
2. S. Chatterjee, S. Bhattacharjee, S.K. Maurya, V. Srinivasan, K. Khare, S. Khandekar, Surface wettability of an atomically heterogeneous system and the resulting intermolecular forces. Europhys. Lett. **118**(6), 68006 (2017)
3. S. Chatterjee, K. Pal Singh, S. Bhattacharjee, Wetting hysteresis of atomically heterogeneous systems created by low energy inert gas ion irradiation on metal surfaces: liquid thin film coverage in the receding mode and surface interaction energies. Appl. Surf. Sci. **470**, 773–782 (2019)
4. A. Chowdhury, S. Bhattacharjee, Experimental investigation of change in sheet resistance and Debye temperatures in metallic thin films due to low-energy ion beam irradiation. J. Phys. D. Appl. Phys. **46**, 435304 (2013)
5. K. Pal Singh, J. Majumdar, S. Bhattacharjee, Tuning optical properties of atomically heterogeneous systems created by plasma-based low-energy ion beams. Appl. Opt. **59**, 4507–4516 (2020)
6. P.B. Johnson, R.W. Christy, Optical constants of the noble metals. Phys. Rev. B **6**, 4370–4379 (1972)
7. H.U. Yang, J. D'Archangel, M.L. Sundheimer, E. Tucker, G.D. Boreman, M.B. Raschke, Optical dielectric function of silver. Phys. Rev. B **91**, 235137 (2015)
8. D. Rioux, S. Vallières, S. Besner, P. Muñoz, E. Mazur, M. Meunier, An analytic model for the dielectric function of Au, Ag, and their alloys. Adv. Opt. Mater. **2**, 176–182 (2014)
9. P.G. Etchegoin, E.C. Le Ru, M. Meyer, An analytic model for the optical properties of gold. J. Chem. Phys. **125**, 164705 (2006)
10. R.L. Olmon, B. Slovick, T.W. Johnson, D. Shelton, S.-H. Oh, G.D. Boreman, et al., Optical dielectric function of gold. Phys. Rev. B **86**, 235147 (2012)
11. W.L. Barnes, A. Dereux, T.W. Ebbesen, Surface Plasmon subwavelength optics. Nature **424**, 824–830 (2003)
12. J. Homola, S.S. Yee, G. Gauglitz, Surface plasmon resonance sensors: review. Sens. Actuators B Chem. **54**, 3–15 (1999)
13. J. Zhang, L. Zhang, W. Xu, Surface plasmon polaritons: physics and applications. J. Phys. D. Appl. Phys. **45**, 113001 (2012)
14. A. Degiron, T.W. Ebbesen, The role of localized surface plasmon modes in the enhanced transmission of periodic subwavelength apertures. J. Opt. A Pure Appl. Opt. **7**, S90–S96 (2005)
15. W. Bekhti, M. Ghamnia, Optical properties of SiO_2 determined by reflection electron energy loss spectroscopy. Catal. Today **89**, 303–306 (2004)
16. R. Todorov, V. Lozanova, P. Knotek, E. Cernošková, and M. Vlček, Microstructure and ellipsometric modelling of the optical properties of very thin silver films for application in plasmonics, Thin Solid Films 628, 22–30 (2017)
17. A. Devos, A. Le Louarn, Strong effect of interband transitions in the picosecond ultrasonics response of metallic thin films. Phys. Rev. B **68**, 045405 (2003)
18. N. Papanikolaou, Optical properties of metallic nanoparticle arrays on a thin metallic film. Phys. Rev. B **75**, 235426 (2007)
19. D.I. Yakubovsky, A.V. Arsenin, Y.V. Stebunov, D.Y. Fedyanin, V.S. Volkov, Optical constants and structural properties of thin gold films. Opt. Express **25**, 25574–25587 (2017)

20. T.W. Ebbesen, H.J. Lezec, H. Ghaemi, T. Thio, P.A. Wolff, Extraordinary optical transmission through sub-wavelength hole arrays. Nature **391**, 667–669 (1998)
21. W.L. Barnes, A. Dereux, T.W. Ebbesen, Surface plasmon sub-wavelength optics. Nature **424**, 824–830 (2003)
22. J. Homola, S.S. Yee, G. Gauglitz, Surface plasmon resonance sensors. Sens. Actuators B **54**, 3–15 (1999)
23. J. Zhang, L. Zhang, W. Xu, Surface plasmon polaritons: physics and applications. J. Phys. D **45**, 113001 (2012)
24. A. Degiron, T. Ebbesen, The role of localized surface plasmon modes in the enhanced transmission of periodic subwavelength apertures. J. Opt. A **7**, S90–S96 (2005)
25. G.G. Martin Dressel, *Electrodynamics of Solids: Optical Properties of Electrons in Matter*, 1st edn. (Cambridge University Press, 2002)
26. H. Kuzmany, *Solid-State Spectroscopy: An Introduction*, 2nd edn. (Springer, Berlin Heidelberg, 2009)
27. A. Mendoza-Galvan, C. Trejo-Cruz, J. Lee, D. Bhattacharyya, J. Metson, P. Evans, U. Pal, Effect of metal-ion doping on the optical properties of nanocrystalline ZnO thin films. J. Appl. Phys. **99**, 014306 (2006)
28. A.N. Oraevsky, I.E. Protsenko, Optical properties of heterogeneous media. Quantum Electron. **31**, 252–256 (2001)
29. S. Zuccon, E. Napolitani, E. Tessarolo, P. Zuppella, A.J. Corso, F. Gerlin, M. Nardello, M.G. Pelizzo, Effects of helium ion bombardment on metallic gold and iridium thin films. Opt. Mater. Express **5**, 176–187 (2015)
30. I. Di Sarcina, M.L. Grilli, F. Menchini, A. Piegari, S. Scaglione, A. Sytchkova, D. Zola, Behavior of optical thin-film materials and coatings under proton and gamma irradiation. Appl. Opt. **53**, A314–A320 (2014)
31. Q. Wei, H. Liu, D. Wang, S.-X. Liu, Degradation in optical reflectance of Al film mirror induced by proton irradiation. Thin Solid Films **519**, 5046–5049 (2011)
32. M. Kumar, T. Jangid, V. Panchal, P. Kumar, A. Pathak, Effect of grazing angle cross-ion irradiation on Ag thin films. Nanoscale Res. Lett. **11**, 454 (2016)
33. A. Axelevitch, B. Gorenstein, G. Golan, Investigation of optical transmission in thin metal films. Phys. Procedia **32**, 1–13 (2012)
34. I.R. Hooper, J.R. Sambles, Some considerations on the transmissivity of thin metal films. Opt. Express **16**, 17258–17267 (2008)
35. M. Nardello, P. Zuppella, V. Polito, A.J. Corso, S. Zuccon, M. Pelizzo, Stability of EUV multilayer coatings to low energy alpha particles bombardment. Opt. Express **21**, 28334–28343 (2013)
36. R. Denton, R. Campbell, S. Tomlin, The determination of the optical constants of thin films from measurements of reflectance and transmittance at normal incidence. J. Phys. D **5**, 852–863 (1972)
37. D. Rioux, S. Vallières, S. Besner, P. Muñoz, E. Mazur, M. Meunier, An analytic model for the dielectric function of Au, Ag, and their alloys. Adv. Opt. Mater. **2**, 176–182 (2014)
38. S. Babar, J. Weaver, Optical constants of Cu, Ag, and Au revisited. Appl. Opt. **54**, 477–481 (2015)
39. P.B. Johnson, R.-W. Christy, Optical constants of the noble metals. Phys. Rev. B **6**, 4370–4379 (1972)
40. M.A. Ordal, R.J. Bell, R.W. Alexander, L.L. Long, M.R. Querry, Optical properties of fourteen metals in the infrared and far infrared: Al, Co, Cu, Au, Fe, Pb, Mo, Ni, Pd, Pt, Ag, Ti, V, and W. Appl. Opt. **24**, 4493–4499 (1985)
41. G.A. Niklasson, C. Granqvist, O. Hunderi, Effective medium models for the optical properties of inhomogeneous materials. Appl. Opt. **20**, 26–30 (1981)
42. A. Sihvola, Mixing rules with complex dielectric coefficients. Subsurf. Sens. Technol. Appl. **1**, 393–415 (2000)
43. M.Y. Koledintseva, R.E. DuBroff, R.W. Schwartz, Maxwell-Garnett model for dielectric mixtures containing conducting particles at optical frequencies. Prog. Electromagn. Res. **63**, 223–242 (2006)

44. B. Segall, Fermi surface and energy bands of copper. Phys. Rev. **125**, 109–122 (1962)
45. H. Ehrenreich, H. Philipp, Optical properties of Ag and Cu. Phys. Rev. **128**, 1622–1629 (1962)
46. A.D. Rakic, Algorithm for the determination of intrinsic optical constants of metal films: application to aluminum. Appl. Opt. **34**, 4755–4767 (1995)
47. A.I. Barreda, J.M. Sanz, R.A. de la Osa, J.M. Saiz, F. Moreno, F. González, G. Videen, Using linear polarization to monitor nanoparticle purity. J. Quant. Spectrosc. Radiat. Transf. **162**, 190–196 (2015)
48. G. Videen, P. Chýlek, Scattering by a composite sphere with an absorbing inclusion and effective medium approximations. Opt. Commun. **158**, 1–6 (1998)
49. J. Sanz, R.A. de la Osa, A. Barreda, J. Saiz, F. González, F. Moreno, Influence of pollutants in the magneto-dielectric response of silicon nanoparticles. Opt. Lett. **39**, 3142–3144 (2014)
50. D. Aspnes, E. Kinsbron, D. Bacon, Optical properties of Au: sample effects. Phys. Rev. B **21**, 3290–3299 (1980)
51. J. Toudert, R. Serna, I. Camps, J. Wojcik, P. Mascher, E. Rebollar, T.A. Ezquerra, Unveiling the far infrared-to-ultraviolet optical properties of bismuth for applications in plasmonics and nanophotonics. J. Phys. Chem. C **121**, 3511–3521 (2017)

Chapter 10
Further Insights on Ion Beam Interaction with Materials: Optical, Electronic, and Space-Related Effects (Case studies 4–7)

In this chapter, we examine ion beam interactions with materials and illustrate their consequences through selected applications. Further case studies highlighting optical, electronic, and space-related effects are presented. The investigations employ materials such as Al, Ag, and Cu, exposed to a broad range of ion beam fluences (10^{15}–10^{18} cm^{-2}) and energies (0.5–3 keV), using Ar, Kr, and He ions.

For the optical studies, the real and imaginary components of the refractive index of both pristine and ion-irradiated samples are determined using the Kramers-Kronig and pseudo-Brewster angle techniques. For mixed media formed as a consequence of irradiation, the Bruggeman and Maxwell-Garnett (MG) effective medium approximations are applied to account for atomic-scale heterogeneity and the optical parameters are thus determined. Additionally, the skin depth of ion-irradiated materials is evaluated across incident light wavelengths corresponding to photon energies between 1 and 5 eV.

In the space-relevant application, films with thicknesses greater than their optical skin depth (ensuring complete opacity), are employed as mirrors. The reflectivity of these materials is measured as a function of both wavelength and ion energy to assess irradiation-induced degradation. The refractive index (real and imaginary parts) is further evaluated using the MG approximation. Notably, incorporating RMS surface roughness through the Debye-Waller formalism leads to a substantial improvement in the agreement between theoretical and experimental reflectivity.

For the electronic application, field emission from ion-irradiated metallic thin films is investigated. A detailed background of the field emission process is provided to contextualize these results. The classical Fowler-Nordheim theory, along with recent relevant theoretical developments, is discussed. Ion irradiation is found to modify the work function due to a combination of subsurface ion implantation and the formation of pointed micro-conical surface structures. Work function values are measured using Kelvin Probe Force Microscopy (KPFM), revealing significant differences between valley regions and ion-induced micro-conical (hillock) features.

S. Bhattacharjee, *Atomically Heterogeneous Systems*,
https://doi.org/10.1007/978-981-95-9733-8_10

An effective work function is then defined, enabling the determination of field emission current and the field enhancement factor. The ion-irradiated films exhibit high emission currents with substantially reduced turn-on voltages, underscoring their potential for applications requiring strong electron emission. A comparative analysis of field emission performance between Ar- and Kr-irradiated samples is also presented. Further details are provided in the sections that follow.

10.1 Optical Effects and Their Applications

The optical constants of a material. Namely refractive index (n) and extinction coefficient (k), play a crucial role in various scientific and industrial applications. These constants define how light interacts with a material, affecting reflection, absorption, and transmission properties. Optical constants provide insights into the electronic and band structure of materials. They help in studying interband transitions, free-electron behavior, and plasmonic effects in metals. A knowledge of the optical constants help in designing anti-reflective coating, mirrors, and optical filters. They are also employed in metallic coatings for solar cells, organic light emitting diodes, and photodetectors to control light absorption and reflection. Optical constants dictate the radiative heat transfer in materials, they are important in thermal coatings and infrared applications, and also for designing nonlinear optical materials for frequency conversion and ultrafast optics. A knowledge of optical constants are crucial for bio-photonic applications such as laser surgery, optical coherence tomography and fluorescence imaging and in optical biosensors. Therefore accurate determination of optical constants is essential for optimizing material performance in diverse fields, from plasmonics and optoelectronics to thermal engineering and biomedical imaging.

Based on the investigations discussed throughout this book, and considering the significance of the topic, numerous techniques have been developed to evaluate the optical constants of pristine metallic thin films, including: (1) investigation on the variation of reflectivity and transmissivity by employing Kramers-Kronig (KK) technique [1–3], and inversion method of Fresnel's equation [4, 5], along with the estimation of the real (n) and imaginary (k) values of the refractive index or vice-versa for materials exhibiting well-defined absorption bands within the considered measurement interval using KK technique. This type of absorption is typically observed in organic materials [6]. Johnson and Christy (JC) provided some of the most reliable measurements available for the optical constants of noble metals [7]. The optical constants of Cu, Ag, and Au under ion beam irradiation of energy 0.1–6 eV using the KK method, was estimated by Babar et al. which was also supported by experimentally measured magnitudes of absorption ($R = 1 - A$) [8]. The results are in a good agreement with those reported by Johnson and Christy as well as other researchers [9–11, 4, 8]. In a similar direction, the research work of Ordal et al. [12] and Palik et al. [13] further contributed to the understanding of optical properties of noble metals. Later on, an universal method for extracting a complex dielectric constant using the experimentally observed optical constants was proposed by

Kuzmenko et al. In this approach, the imaginary part is parameterized independently at each frequency point, while the real part is obtained through the KK transformation [14]. Grosse et al. [15] extended the KK technique to handle oblique incidence and multilayered systems, demonstrating its applicability using the reflectance spectra of BaF_2 and NaCl to study their multiphonon behavior. Palmer et al. [2] applied KK analysis to the transmittance spectra of a single thin film, evaluating the algorithm on a CO_2 film deposited on a 20 K Ge substrate. The KK formalism has consequently become a cornerstone for extracting the optical constants of metals, semiconductors, and dielectric materials [1, 2, 16]. Nevertheless, the methodologies discussed above are primarily suited for optically homogeneous media, such as pristine metallic thin films (MTFs). When dealing with heterogeneous or composite systems, such as ion-modified or nanostructured films where additional theoretical frameworks are required, the effective optical response of such mixed phases, has employed several effective medium theories, including the Maxwell-Garnett (MG) model, the Bruggeman formulation, and related effective medium approximations [17–20]. In our earlier work, both the MG and Bruggeman models were evaluated for ion-irradiated Cu, Ag, Al, and Au films, where the MG approximation demonstrated superior consistency with experimental refractive index (n) and extinction coefficient (k) values determined using the pseudo-Brewster angle technique [21]. Accordingly, the MG framework has been adopted here as a reference for assessing the experimental trends in the real (n) and imaginary (k) parts of the refractive index, as functions of ion fluence and photon energy. Beyond refractive-index extraction, additional optical parameters, including skin depth and optical conductivity have been systematically evaluated for Ag, Al, and Cu thin films subjected to low-energy ion irradiation over fluences in the range of $(1.6–7.2) \times 10^{15}$ cm^{-2}.

A detailed experimental investigation on the optical constants of unbombarded metallic thin films are determined using the KK approach, which can be found in Ref. [22]. In summary, the method begins with recording the reflectance spectra of the as-deposited films through UV-VIS-NIR spectrophotometric measurements. These spectra are then processed using the KK relations, which allow the wavelength-dependent refractive index and extinction coefficient to be derived over the 250–1200 nm spectral window [21, 22]. For metallic layers exposed to ion irradiation, their optical parameters—such as the refractive index, extinction coefficient, skin depth, optical conductivity, and dielectric response—were determined within the framework of the MG effective medium model. The accuracy of the n and k values predicted by MG was assessed by comparing them with experimentally obtained data from pseudo-Brewster angle measurements carried out at four laser wavelengths (405 nm, 532 nm, 632.8 nm, and 670 nm). The strong agreement between the two sets of results validates the MG-based analysis. Silver and aluminum thin films, along with the benchmark copper system, were selected for comprehensive characterization because of their significant technological importance [21]. These materials are widely integrated into devices such as solar energy conversion systems [23, 24] and biosensing architectures [25, 26]. Additionally, Ag and Al coatings are commonly employed in high-reflectivity components for satellites and spaceborne antenna systems, making it crucial to understand how their optical behavior evolves under exposure to ionizing radiation.

10.1.1 Kramers-Kronig Technique and Its Application to Homogeneous Media

In the Kramers-Kronig (KK) technique the phase shift of the reflected light to the incident light is determined. The obtained phase information along with the Fresnel's equation can be employed to provide the optical constants of the medium. When light strikes an air ($n_1 = 1$)—metal ($n_2 = n + ik$) interface, the reflection coefficient can be given by the following equation [1] .

$$\tilde{r}(\omega) = \rho(\omega) \exp i\theta(\omega) = \frac{1 - (n + ik)}{1 + (n + ik)}, \tag{10.1}$$

where $\rho(\omega)$ is the reflectivity modulus and $\theta(\omega)$ is the phase difference between the incident and reflected light, n and k are the real and imginary parts of the refractive index, and the reflectivity R is defined by the following relation:

$$R = |\tilde{r}(\omega)|^2 = \rho(\omega)^2. \tag{10.2}$$

The following relation between the reflection coefficient $\tilde{r}(\omega)$, reflectivity R, and the phase difference $\theta(\omega)$ between the incident and reflected light can be obtained by taking the natural logarithm of Eq. 10.1 and employing Eq. 10.2,

$$\ln r(\omega) = \ln \sqrt{R} + i\theta(\omega). \tag{10.3}$$

Furthermore, the phase difference between the incident and reflected light beam at a particular frequency ω_o can be obtained by employing the KK relations [1] for $\ln r(\omega)$ as follows:

$$\theta(\omega_o) = -\frac{2\omega_o}{\pi} \int_0^{\infty} \frac{\ln \sqrt{R}}{\left(\omega^2 - \omega_o^2\right)} d\omega = \frac{1}{2\pi} \int_0^{\infty} \ln\left|\frac{\omega - \omega_o}{\omega + \omega_o}\right| \frac{d(\ln R(\omega))}{d\omega} d\omega. \tag{10.4}$$

The phase function $\theta(\omega)$ can be obtained by evaluating Eq. 10.4 over a range of frequencies ω. Although the KK relation formally requires integration from zero to infinite frequency to recover the exact phase, such an extensive spectral measurement is experimentally impractical. Consequently, it is well established that the reflectance spectrum must be reasonably extrapolated beyond the measured window to ensure accurate phase reconstruction using the KK formalism [8, 27]. In Eq. 10.4, the first logarithmic term on the right-hand side approaches zero in the limits $\omega \gg \omega_0$ and $\omega \ll \omega_0$ because $\ln(1) = 0$. Moreover, if the reflectance $R(\omega)$ varies only weakly with frequency over a given region, the second term of the integrand becomes negligible. These considerations imply that direct reflectivity measurements across an infinite spectral span are unnecessary when the objective is to analyze the phase behavior associated with a particular resonance at ω_0. In this work, reflectance spectra for all metallic thin films were recorded using a UV-VIS-NIR spectrophotometer over the photon-energy range of ~ 1–4.96 eV (250–1200 nm) with a spectral step of 2 nm [21]. To satisfy the KK integration requirements, $R(\omega)$ for Cu, Ag, and Al was extended to lower and higher photon energies, down to 0.8 eV and up to 10 eV, using published reflectivity datasets for copper [28], silver [28], and aluminum [27]. From Eqs. 10.1–10.3, the optical constants namely n and k of the refractive index of the material can be readily derived as a function of the reflectivity R and phase angle θ as follows,

$$n = \frac{1 - R}{1 + R + 2\sqrt{R}\cos\theta}, \tag{10.5}$$

$$k = \frac{2\sqrt{R}\sin\theta}{1 + R + 2\sqrt{R}\cos\theta}. \tag{10.6}$$

The refractive index (n) and extinction coefficient (k) are determined directly from the experimentally measured reflectance R and the phase $\theta(\omega)$ obtained through the KK transformation in Eq. 10.4. However, applying these relations requires careful consideration when dealing with materials that are not optically uniform. Equations 10.5 and 10.6 are strictly valid only for media that are linear, isotropic, and homogeneous in nature [18]; hence, they cannot be reliably used for systems exhibiting inhomogeneity. For this reason, KK-based extraction of optical constants as a function of photon energy has been employed exclusively for pristine metallic thin films in the present study.

For metallic thin films subjected to energetic ion-beam exposure, the resulting microstructural modifications introduce heterogeneity into the medium, rendering the direct application of KK-derived relations inappropriate. In such cases, the Maxwell-Garnett (MG) effective medium approximation is employed to estimate the modified dielectric response, using the pristine dielectric constants extracted from the KK analysis as baseline input. Effective medium theories, including the MG model, are well established for capturing irradiation-induced changes in optical properties and have been shown to accurately describe the evolution of dielectric behavior in ion-modified metallic films [21, 19].

Figures 10.1 and 10.2 present the extracted values of the refractive index (n), extinction coefficient (k), and skin depth (δ) for aluminum, copper, and silver thin films as functions of incident beam energy and ion fluence. The results are shown for three representative fluence levels: 1.6×10^{15} cm^{-2}, 5.0×10^{15} cm^{-2}, and 7.2×10^{15} cm^{-2} with a variation of ion beam energy from 1 and 5 keV.

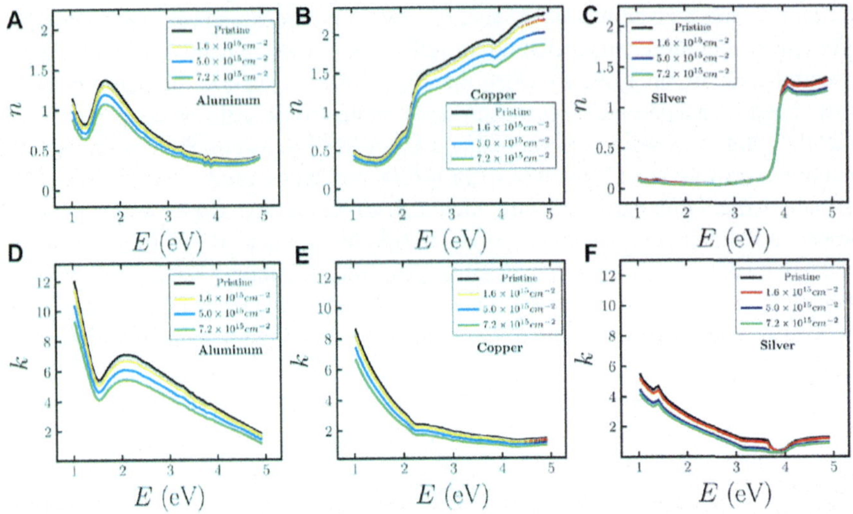

Fig. 10.1 Variation of real and imaginary parts of the refractive index for aluminum (**a**, **d**), copper (**b**, **e**), and silver (**c**, **f**), respectively, with energy of incident light as well as varying fluence of Argon ions. Black solid line represents the data obtained from KK technique and colored (red, blue, and green) solid lines represent the data obtained from MG approximation. Reprinted from [22] Frontiers, under CC-BY license [29]

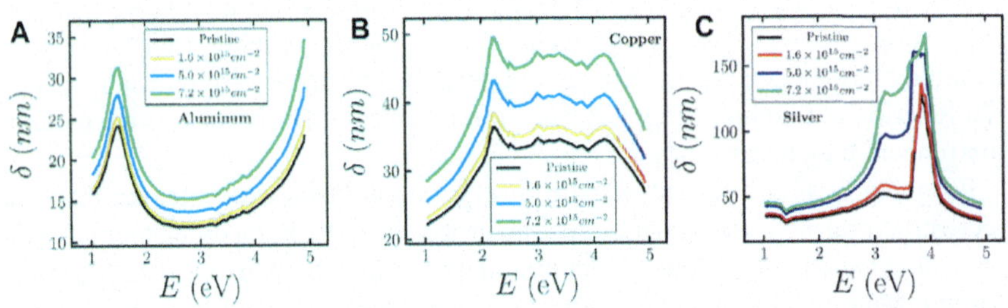

Fig. 10.2 Variation of skin depth for **a** aluminum, **b** copper, and **c** silver, respectively, with energy of incident light as well as varying fluence of Argon ions. Reprinted from [22] Frontiers, under CC-BY license [29]

10.2 Effect of Ion Beam Irradiation on Space Relevant Materials and Their Applications

Space is a hostile environment consisting of high-energy charged particles which can be ionizing including non-ionizing radiation. Spacecrafts are therefore constantly bombarded with such types of radiation [30]. It is of utmost importance to investigate the sustainability of space relevant components which are employed in spacecrafts, on whether they would survive the ionizing and non-ionizing radiation. Therefore, often it is felt advantageous to perform controlled experiments in the laboratory where the space relevant components can be exposed to the right amount of dose that the component would experience if it were in space for a certain duration of time. At first, it may be worthwhile to look at the different types of radiation that a satellite is exposed in a space environment. There are trapped particles coming from the radiation belts in the Earth's magnetosphere, the solar particles coming from the sun and lastly the galactic cosmic radiation (GCR) coming from the galaxy [30, 31]. The sun emits radiation in the form of solar wind, which are mainly protons and electrons having a wide range of energies ranging from about 100 eV to about 3.5 keV for protons. Particles that are trapped in the Earth's radiation belts consist of electrons of up to about 10 MeV, protons of up to about 700 MeV, and heavier ions of less than about 50 MeV/nucleon. Sometimes considerable amounts of neutrons can be encountered in Low Earth Orbit (< 1000 km altitude). They are usually neutrons that are produced in the Earth's atmosphere where GCR interact with air molecules.

As a result of bombardment by charged particles in space, the performance of optical components used in space vehicles is subjected to degradation in the harsh space environment [32, 33]. Therefore, if prior knowledge can be attained through controlled experiments in the laboratory, then it would be quite helpful to assess the amount of such damage brought about by the space radiation effects of ions such as helium and hydrogen on space relevant materials such as Al and Ag metallic thin films (MTFs). Then a quantitative judgment can be made regarding the possibility of using such materials in the space environment for making robust metallic film mirrors that can sustain the ionic bombardment in space. As mentioned above, protons (H^+) which are the most abundant in the solar wind, making about 95% of it, and alpha particles (He^{2+}) which are doubly ionized helium and are the second most abundant in the solar wind making about 4% of it, including cosmic rays, can bombard the metallic thin films used in spacecrafts and satellites and disturb the proper working of the components made with such thin films [34].

There are several research works on the implantation effect of high-energy protons [35, 36] in the literature, however, studies on the irradiation effect of low-energy protons and alpha particles are relatively few. There are reports suggesting that the typical kinetic energy lies in the range of ~ 0.4–7.5 keV for protons and ~ 1.6–30 keV for alpha particles [37]. In addition to that mentioned above, Coronal Mass Ejection (CME) and solar flares from the sun can lead to very energetic solar wind particles with energies of the order of hundreds of keV or in MeV [38].

The present research is aligned with the objectives of the European Space Agency's Solar Orbiter (SOLO) mission, designed to investigate solar phenomena by approaching the Sun as close as 0.28 AU at perihelion [38]. Scientific findings from SOLO have shown that optical elements onboard spacecraft undergo significant degradation, primarily due to sustained exposure to both low- and high-energy protons, alpha particles, and electrons [30, 38].

To contextualize the present study, key research efforts on ion-beam irradiation of space-relevant materials are briefly summarized. Pelizzo et al. [30] examined the degradation of Si/Mo and Ir/Si multilayer optical coatings subjected to 1 keV proton bombardment, simulating exposure equivalent to 3 months and 1 year of the trajectory of the Solar Orbiter. These multilayer stacks are integral to multi-element telescopes and imaging spectrometers used in the mission [39, 40]. Similar degradation phenomena have also been reported for ultraviolet photolithography optics, where low-energy proton exposure leads to significant performance loss [41]. Nardello et al. [38] further investigated Si/Mo and Ir/Si multilayer coatings irradiated with 4 keV alpha particles representative of solar wind conditions, with fluences corresponding to one, two, and four years of Solar Orbiter operation. In addition to multilayer systems, numerous studies have focused on ion-irradiation effects in metallic thin films, which are widely used as reflective coatings in space-borne instruments [21, 22, 42–45]. In line with this direction, our earlier laboratory simulations assessed the stability of aluminum and silver thin-film mirrors under low-energy helium ion implantation, offering insights into their durability in solar-wind-like environments [46]. Comprehensive experimental results are available in our previous publication [46]. Complementary work by Zuccon et al. [47] demonstrated that 4 keV helium ions equivalent to the alpha-particle flux encountered over one to four years of solar-wind exposure significantly alter the ultraviolet and visible reflectivity of gold and iridium coatings.

The degradation in reflectivity of aluminum (Al) and silver (Ag) metallic thin films under irradiation by low-energy helium ions, with ion energies of 0.5, 1, 2, and 3 keV and fluences in the range of 1.10×10^{16} cm^{-2} to 1.56×10^{16} cm^{-2} is investigated [48]. The magnitude of these fluences were selected to emulate cumulative ion doses corresponding to approximately four (1.1×10^{16} cm^{-2}) and six (1.56×10^{16} cm^{-2}) years of Solar Orbiter exposure [38], assuming no additional solar events alter the particle flux during this period. The high ion flux achievable in our setup arises from the confinement of helium plasma within a compact multicusp source, allowing the required doses to be delivered in only a few seconds (≤ 3 s). This dramatically reduces the irradiation time compared to other ion-beam facilities used in similar space-environment simulations [30, 38]. The delivered fluence is controlled by the irradiation duration, with an estimated uncertainty of ~ 10%, consistent with our earlier measurements [46, 47]. The metallic thin films, each ~330 nm thick, were designed to fully suppress optical transmission, as this thickness significantly exceeds the optical skin depth. As a result, absorptivity could be directly computed from $A = 1 - R$ using measured reflectance data. Samples of Al and Ag with dimensions ~ 1.25×1.25 cm^2 were uniformly irradiated using an ion beam of ~ 2 cm diameter. Reflectance spectra were recorded before and after irradiation using a L950 UV-VIS-NIR spectrophotometer at an incidence angle of 8°, covering wavelengths from 200 to 2500 nm with a spectral resolution of 5 nm, thus spanning the ultraviolet (UV), visible, and near-infrared (NIR) regions. Key instrument specifications include a 0.5×1 cm^2 entrance slit and operational wavelength coverage from 175 to 3300 nm, with a resolution range of 0.05–5.00 nm in the UV/VIS and 0.20–20.00 nm in the NIR. The reflectivity uncertainty, based on manufacturer specifications, is approximately 0.2–0.3%. A comprehensive presentation of the experimental results is available in our earlier publication [46].

Figure 10.3a–d illustrates the reflectivity spectra of aluminum and silver thin films over the wavelength range of 200–2500 nm for two representative helium-ion fluences 1.1×10^{16} cm^{-2} and 1.56×10^{16} cm^{-2} and for ion energies between 0.5 and 3.0 keV. The reflectance profiles of the two metals exhibit distinctly different spectral characteristics. For aluminum, a pronounced dip consistently appears near ~814 nm in both pristine and irradiated samples, as highlighted in the insets of Fig. 10.3a and b. This feature is associated with the well-known interband transition in aluminum [48, 49], arising from electronic excitation from filled valence-band states to empty conduction-band states near the Brillouin-zone boundary. Importantly, the position of this transition ($\approx$ 814 nm, or $\approx$ 1.5 eV) remains essentially unchanged with variations in ion-beam energy or fluence, for both the four-year and six-year equivalent exposure doses. Silver exhibits an analogous behavior: a characteristic dip appears around ~320 nm, corresponding to the material's interband transition [48], as shown in Fig. 10.3c and d. Similar to aluminum, the spectral position of this transition does not shift with helium-ion fluence or beam energy, indicating that these intrinsic electronic transitions are not significantly perturbed under the irradiation conditions explored. The broader reflectivity trends show that silver maintains an almost constant reflectance beyond ~ 500 nm, with minimal dependence on wavelength or ion-beam parameters. Aluminum, however, reaches a similar saturation only at much longer wavelengths, near ~ 1500 nm. The behavior in the near-infrared (NIR) region (1200–2500 nm) suggests that aluminum thin-film mirrors may be particularly well-suited for reflecting infrared solar radiation without substantial degradation during the four- or six-year Solar Orbiter mission period. Minor modifications in reflectivity with ion-beam energy are observed, but these effects remain relatively small and are more noticeable at shorter wavelengths.

The key factor that plays a crucial role in modulating the reflectivity of ion-irradiated metallic thin films, is RMS roughness. For the 4-year and 6-year equivalent doses of helium ions, the variation in surface roughness remains nearly identical across both scanned areas 20×20 μm^2 and 5×5 μm^2. The corresponding RMS roughness values lie in the range of ~ 19–23 nm and ~ 19–26 nm for the 4-year and 6-year doses, respectively. Among all the examined combinations of irradiation energy and fluence, the samples irradiated at 2 keV with a 6-year equivalent dose exhibit the highest roughness, which plausibly explains the pronounced change in reflectivity observed under these conditions. In contrast, MTF of Ag shows an increase in reflectivity after helium ion irradiation, which can be attributed to a reduction in RMS surface roughness an effect similar to that reported for aluminum thin films. This roughness-driven enhancement in reflectivity is consistent with earlier studies [46, 47]. As previously noted for aluminum, a decrease in roughness is typically associated with grain growth induced by ion irradiation, which may also be the underlying mechanism in the present case [47]. The RMS roughness of the silver MTFs is found to be significantly lower than that of the aluminum films, lying in the range of ~ 1–5 nm and ~ 3–5 nm for the 4-year and 6-year doses, respectively.

Metallic films of thickness $\sim$ 330 nm are prepared for the present experiments, which is significantly greater than the corresponding skin depth, thereby ensuring that the films remain completely opaque. Both the real (n) and imaginary (k) parts of the

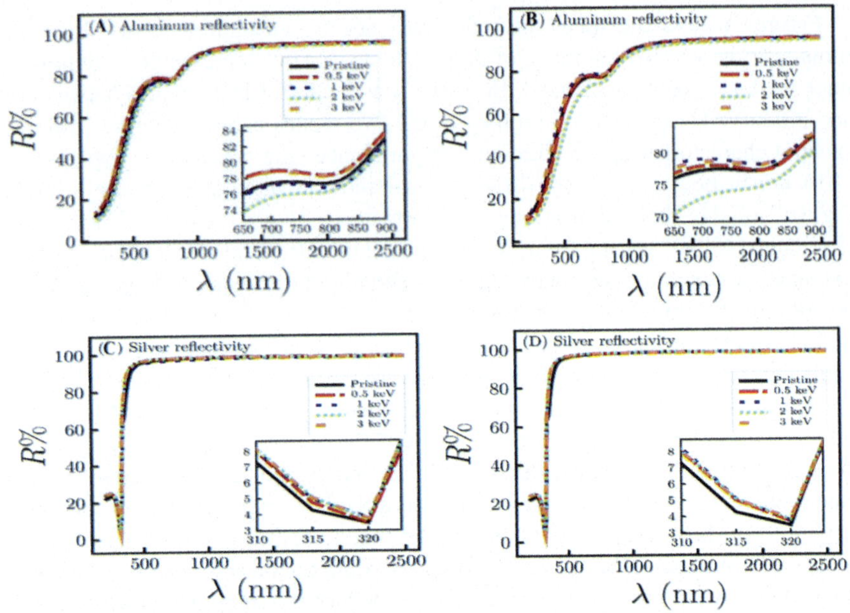

Fig. 10.3 Reflectivity of aluminum (**a** and **b**) and silver metallic thin films (**c** and **d**) irradiated with helium ions with varying energy and at 4 years dose (1.1×10^{16} cm^{-2}) (**a** and **c**) and at 6 years dose (1.56×10^{16} cm^{-2}) (**b** and **d**), as a function of wavelength of incident light. Reprinted with permission from [46] © Optica Publishing Group

refractive index obtained through the MG approximation are found to decrease with increasing He-ion fluence, leading to a reduction in the reflectivity of the metallic medium (cf. Eq. (9.4) of Chap. 9). This behavior can be rationalized by simultaneously considering the irradiation-induced modifications in the refractive index and the RMS surface roughness of the samples.

It is further observed that the agreement between the theoretical and experimental reflectivity improves markedly when the RMS surface roughness is incorporated into Eq. (9.4) of Chap. 9 using the Debye-Waller (DW) formalism [50] provided below. The inclusion of the DW factor effectively accounts for surface-scattering losses arising from roughness, thereby yielding a more accurate description of the reflectivity of ion-irradiated metallic thin films.

$$R = R_o \exp\left(\frac{-8\pi^2\sigma^2 n^2 \cos^2\theta}{\lambda^2}\right), \tag{10.7}$$

Here R_o is the reflectivity obtained from Eq. (9.4) of Chap. 9, σ is the RMS surface roughness in nm, θ is the incidence angle from normal, and λ is the wavelength of the incident light in nm. The measured reflectivity exhibits a strong correlation with the RMS surface roughness particularly in the UV region. This is expected because, at shorter wavelengths, light scattering from a rough surface becomes more

Table 10.1 Magnitude of fractional volume for different ion energies and fluences

Ion energy (keV)	Fluence ($\times 10^{16}$ cm^{-2})	Values of fractional volume f
0.5	1.10	0.20
	1.56	0.29
2.0	1.10	0.05
	1.56	0.07

significant—scattering increases sharply when the wavelength approaches the characteristic size of surface irregularities. Consequently, even small variations in roughness produce noticeable changes in reflectivity in the UV range, whereas the effect diminishes at longer wavelengths.

As the ion fluence increases, the total number of helium ions bombarded on the sample per unit area also increases, leading to an increment of the density and hence fractional volume f of the implanted He atoms. The value of f is evaluated and subsequently used in the MG approximation to determine the refractive index of aluminum, although it has been experimentally estimated via the pseudo-Brewster (PB) angle technique. The values of f for different ion energies and fluences are listed below in Table 10.1. A reasonable agreement is observed between the n and k values obtained from the MG approximation and those obtained from the PB method. The magnitudes of reflectivity , predicted by both MG and PB techniques, are consistent with each other but exceed the experimentally measured reflectivity by approximately 20–25%. However, when the Debye-Waller (DW) correction (Eq. 10.7) is applied, the correspondence between the theoretical and experimental reflectivity improves significantly, reducing the discrepancy to within 9–10%.

10.3 Effect of Ion Beam Irradiation on Metal Surfaces for Field Emission and Their Applications

When a strong electric field is applied to any material such as a metal, the electric field attracts the electrons from the metal and extracts them out of the surface. The electrons are emitted into the vacuum from the metal surface and is known as field emission, which is strongly dependent on the work function of the metal, upon which the field is applied. The process of field emission was explained on the basis of quantum tunneling of electrons in the late 1920s and the theory was proposed in 1928 by Fowler and Nordhiem through an equation after their name. Field emission has many applications, for example, as bright electron sources for high-resolution electron microscopes, field emission displays, electron beam lithography, liquid metal ion source-based focused ion beams, X-ray generation, microwave tubes, vacuum electronics and surface analysis. Field emitted electrons from materials that have a low turn on voltage and high current densities find applications in areas such as field emission cathodes (vacuum electronic devices), gas sensors, magnetic storage devices, flat panel displays, and microwave amplifiers including catalytic reactors [51–55]. Conventionally, field emission has been studied in flat and homogeneous metal surfaces. However, there are many studies which have shown that field emission from conical and pointed micro-structures leads to a large enhancement of the electric field and profuse generation of the field emission current. Usually, such field emitter arrays have been prepared on a silicon substrate by lithographic techniques. The original field emitter array was a Spindt array in which the individual field emitters were made up of sharp molybdenum cones. The technology was developed at SRI International by C. A. Spindt and the first article on a single such emitter was published in 1968. Due to the pointed tips even a small gate voltage (< 100 V) could lead to the development of a high local electric field. Therefore, the physical texturing of a material surface into micro-conical structures seemed to be an attractive option for field emission applications. While development of Spindt array is an involved process, we have explored an easier method on the generation of arrays of micro-metallic field emitters by employing low-energy ion beam irradiation.

Exposure of material surface to the energetic ion beam irradiation offers several advantages, such as fabrication of contamination-free nano and microstructures with controllable dimensions, that can be tailored via geometry, composition, and concentration [56–58]. Inert gaseous ions such as He and Ar are widely employed to create diverse surface morphologies including fuzz, craters, dots, and ripple patterns on carbon nanotubes, graphene sheets, and metallic or semiconducting materials such as ZnO, Si, Ge, and high-Z metals like W, Mo, Ta, and Pd [59–62]. These type of structures are mostly self-organized, and the associated increase in effective surface area due to pattern formation results in naturally emerging sharp tips that are well suited for field emission applications. The field emission performance of such structures is strongly influenced by the tip geometry, aspect ratio (height-to-base width), areal density of emission sites, and local variations in work function. Consequently, researchers have explored multiple strategies to enhance field emission characteristics, including increasing the carrier concentration via ion doping [63], generating microstructures through plasma processing [64, 65], and producing ordered surface patterns through ion bombardment [66].

Field emission phenomena can manifest in both advantageous and detrimental ways, depending on the application. Several investigations have shown that ion bombardment of metallic surfaces can cause impure atoms, particularly those with low sputter yields originating from the substrate, to redeposit onto the surface, thereby giving rise to a disordered landscape of conical protrusions commonly referred to as sputter cones [67–69]. Conversely, in systems where metallic components operate under extremely high electric fields (exceeding the MV/m range), such as plasma-facing materials in fusion devices or high-gradient regions of particle accelerators, field-emitted electrons can ionize background gas species. This process may trigger an avalanche mechanism that culminates in electrical breakdown [70]. Consequently, significant effort has been devoted to uncovering the fundamental processes that govern electron emission from metal surfaces subjected to such extreme field conditions [71, 72]. Beyond morphological evolution, ion implantation can also induce nanoscale compositional variations. The incorporation of irradiated ions within shallow interstitial layers introduces atomic-level heterogeneity that further modifies surface electronic properties and emission behavior [44].

In our laboratory, we have undertaken systematic investigations on ion beam-irradiated metallic surfaces, where the simultaneous creation of microscale surface features and atomic-level heterogeneity significantly modifies the field emission behavior. In addition to the geometric enhancement provided by ion-induced protrusions, the incorporation of implanted species alters the local electronic structure, giving rise to an effective work function that reflects the combined contributions of the host metal and the embedded impurity atoms. Such modifications can substantially influence electron emission characteristics. A number of studies employing low-energy ion beams have demonstrated pronounced changes in the electrical, surface, and optical response of metals following the development of these atomically heterogeneous layers [21, 44–73]. It is also well established that low-energy ion bombardment can generate ordered or quasi-ordered micro-protrusion arrays, which act as efficient emission sites due to their enhanced local electric-field concentration [74, 75]. Copper serves as an excellent model system for these studies. It is abundant, cost-effective, and possesses an electrical conductivity ($\sigma_{Cu} = 5.96 \times 10^7$ S·m^{-1} at 20 °C) comparable to that of noble metals such as Ag and Au. Furthermore, Cu is widely used across a diverse range of technologies, including sensing devices, aerospace components, and photocatalytic systems, making it an attractive material for exploring ion beam-mediated tuning of functional properties [76–78].

Irrespective of extensive investigations on plasma-induced surface modification, systematic field-emission investigations on copper substrates subjected to low-energy ion irradiation remain scarce. In our work, we examine the underlying pathways through which self-organized surface features emerge under such irradiation conditions, and we elucidate how these structures contribute to enhanced electron emission. Notably, the conical protrusions formed on Cu surfaces in our experiments arise without the involvement of secondary metallic impurities, distinguishing them from the sputter-cone structures commonly reported in impurity-driven growth. The field-emission characteristics of these impurity-free, ion-induced conical architectures, particularly their response under high electric fields, have not been previously documented, highlighting the novelty and significance of the present study.

A wide range of industrial technologies increasingly demand large-area, micro- to nanoscale surface texturing, and low-energy ion irradiation (0.1–3 keV) has emerged as a particularly versatile route for achieving such pattern formation [79]. By tuning the incident ion energy and fluence, one can precisely regulate the implantation depth and the density of ions delivered to the surface, enabling controlled modification of near-surface atomic configurations [80]. The concept of self-organized pattern formation under ion bombardment dates back to the pioneering observations of Navez et al. in 1962 [81], which demonstrated that energetic ions can spontaneously reorganize surface atoms into ordered morphologies. Since then, ion-irradiation-driven self-assembly has evolved into a major research area, with numerous studies reporting the formation of nanodots [82], ripples [83–85], fuzz layers [86], bubble-like features [87], and a variety of related architectures [88] on metallic substrates including Au, Ag, W, and Mo, as well as on semiconductors such as Si and Ge [89], and their alloys under energetic ion exposure [90–92]. Parallel efforts have shown that carbon-based materials, ranging from carbon nanotubes and diamond films to graphitic structures like graphene also undergo rich surface pattern evolution under ion irradiation [93–95], further underscoring the universality and technological potential of ion-induced self-organization.

The theoretical aspect of the mechanism that governs the evolution of surface morphology under ion irradiation, have been extensively examined by several groups [96–98], and a brief overview of these models is provided in the following section. Given the broad technological importance of field emission [94, 95, 99–105], considerable effort has also been directed toward establishing a fundamental theoretical framework for emission from ion-modified surfaces. When inert gas ions are used as projectiles, the resulting atomically heterogeneous near-surface layers preserve the overall chemical composition of the parent material [106, 107], while introducing nanoscale disorder and implanted species. As highlighted earlier, such structural perturbations—both in terms of modified topography and embedded ions—directly influence the local work function and field-enhancement factor. Consequently, low-energy ion irradiation provides a powerful and catalyst-free approach for tailoring the spatial distribution and self-organization of field-emitting sites through controlled adjustment of ion flux, ion energy, and angle of incidence [108–112]. In our work, we have further examined the evolution of these ion-induced surface structures and evaluated existing theoretical descriptions of their growth dynamics [113–115].

As noted earlier, copper was selected as the substrate owing to its attractive combination of physical and chemical properties [116–118, 74, 78]. Its high electrical conductivity, favourable electronic structure, and earth abundance make it an appealing alternative to costly noble metals for applications spanning catalytic organic transformations, electrocatalysis, and photocatalysis. These features position Cu as a versatile and economically viable platform for ion-beam-induced surface engineering. The detailed experimental outcomes of our study are reported in Ref. [107].

10.3.1 The Process of Field Emission: A Background on Existing Theories

A variety of theoretical models describe the growth phenomena as competitive processes between orientation-dependent surface diffusion process and sputtering. The Ehrlich-Schwoebel model describes interlayer diffusion in epitaxial systems, where it talks about a barrier that describes how atoms diffuse on a surface when they encounter an atomic step and has been employed to study surface morphology and growth in the fabrication of electronic devices [119]. The model of Martynenko and Nagel describes the process of diffusion of adatoms, their clustering, and the cluster growth resulting in polycrystal formation [98]. A theory of ripple topography arising due to off-normal incidence ion beam was proposed by Bradley-Harper [96].

Cuerno and Barabasi derived a stochastic nonlinear equation to describe the evolution and scaling properties of surfaces eroded by ion bombardment [97]. The ripple formation on ion-irradiated Si substrate has been explained from the point of view of sputtering of concave surfaces and deposition happening on convex surfaces, which becomes more pronounced when irradiation is undertaken at an angle. There are reports where impure metal atoms rest on conical protrusion generated with ion beam irradiation and protect the underlying surface atoms from getting eroded by the incoming flux of ions [113, 120]. Robinson and Rossnagel's enhanced diffusion model deals with the role of ion diffusion which cannot be only explained with thermal diffusion effects alone but with enhanced diffusion due to cooperative effects between multiple ion impacts [80]. Davidovitch et al. showed theoretically that there are certain response functions that make the flat surface linearly stable while subtly altering the predicted sputter yield as a function of incidence angle. There is a window of stability in ion bombarded surfaces across a range of beam angles for specific ion and surface types. Additionally, they describe key characteristics of pattern evolution near the transition from unstable to stable dynamics [114]. An early time dynamics model was proposed by Bradley where it was shown that an initial flat solid surface can be destabilized if the impure ionic flux exceeds a certain critical value [115]. Perkinson proposed that morphological evolution of ion-bombarded surfaces in terms of average surface height response to individual impacts which has been generalized using the crater function formalism often do not agree with precise experimental measurements employing low-energy inert gaseous ions in the 1 keV range acted upon by Si or Ge as substrates [121]. A central feature of our experimental approach is that no external impurity species is introduced during irradiation; consequently, the development of surface structures cannot be attributed to differential sputtering effects. To rationalize the emergence of conical protrusions, we propose that each ion impact locally removes the material and subsequently redistributes it through collision-induced rearrangement. This redistributed material aggregates into clusters of substrate atoms, which then serve as nucleation centres that promote and sustain the growth of the protrusions.

The Fowler-Nordheim electron emission induced by an electrostatic field, described a process that happens through quantum tunneling of electrons through a surface-vacuum interface in the presence of a high electric field (> 1 GV/m). The process named after the discoverers Fowler and Nordheim can be described by the equation below, that assumes a smooth planar metal surface with a given work function, free electron model (Sommerfeld), and a semi-classical approximation (WKB) for the tunneling probability. The current density J from the Fowler-Nordheim (FN) equation is given by the following equation [122, 123]:

$$J = \frac{A(\beta E)^2}{\phi^{3/2}} \exp\left(-\frac{B\phi^{3/2}}{E\beta}\right), \tag{10.8}$$

where E is the applied electric field, ϕ is the bulk work function, $A = 1.54 \times 10^{-6}$ AV^{-2} eV, and $B = 6.83 \times 10^{7}$ V cm^{-1} eV$^{-3/2}$ are constants. The field enhancement factor (β) is a morphology and work function-dependent parameter, affecting the local field around micro (nano)-scale protrusions on the metal surface. As seen from the FN equation (Eq. 10.8), the field emission current is a function of ϕ and β, which in turn can be modified by the aspect ratio of the developed structures and the embedded ionic impurities on the surface that can change the local work function. The curvature of the nanoscopic emitter was studied by Kyritsakis et al., [124] and a generalized FN-like equation was derived. Similarly, Biswas et al., looked at the altered tunneling potential and concluded that curvature correction is essential for apex radius less than a micron [125]. Djurabekova et al. [108] employed Density Functional theory (DFT) calculations and analyzed the redistribution of electron density on the copper (100) surface due to self-adatoms and in the presence of high electric fields. The modification to the work function around rough surface features and microstructures were also investigated. Toijala et al. [126] performed density functional theory and quantum transport calculations to study the effect of atomic scale defects on the work function and field emission properties of metal surfaces, particularly the copper (111) surface. However, from this model, it is difficult to justify the effective high value of the field enhancement factor in the field emission characteristics of a metal surface, as observed in their as well as our experiments.

The setup used for carrying out field emission measurements, is presented in Chap. 5 and detailed further in Refs. [25, 107]. Measurements were performed on both pristine copper substrates and samples irradiated to produce well-defined micro-conical structures. In the extraction stage, ions generated in the microwave-sustained plasma are accelerated through the metallic electrode system. All ions reach the plasma electrode aperture (PLE) with approximately the Bohm velocity and carry energies of order eV_p, where V_p (typically a few tens of volts) is the plasma potential relative to ground. Downstream of the PLE, the ions are further accelerated by the applied collector bias ($V_c \sim$ kV), enabling their extraction into the high-vacuum region of the ion extraction chamber [127]. Since the ions are extracted directly into a high-vacuum region, they experience negligible ion neutral scattering due to the effective differential pumping employed downstream of the plasma chamber. Consequently, the extracted beam can be treated as essentially monoenergetic, with the ion energy governed primarily by the applied collector potential (V_c), as the contribution from the plasma potential (V_p) is comparatively small (the corresponding energies, $eV_p \ll eV_c$). Under typical operating conditions, the discharge is dominated by singly charged ions, an assumption consistent with earlier studies of similar plasma sources [128]. Details of sample fabrication are provided in Chap. 7, while the associated characterization procedures are described in Chap. 5.

Surface modification via ion beam irradiation were conducted over a range of ion fluence from 1.3×10^{18} to 4.7×10^{18} cm^{-2}. Starting from the generation of plasma, the uncertainity in estimating ion fluence is found to be approximately 4.2%, consistent with standard diagnostics used for plasma-based ion extraction. A beam diameter of 1 cm was maintained throughout the study to ensure spatially uniform irradiation across the exposed region of each sample. Following irradiation, all specimens were stored in vacuum desiccators to avoid post-treatment surface contamination from ambient atmospheric exposure.

During the experiments, both pristine and ion-irradiated samples exhibiting micro-cone formation, under ambient (room temperature) conditions were used without any external heating. The heat generated during ion bombardment is effectively dissipated via the conducting as well as radiating sample holder made of copper (high thermal conductivity). To further eliminate any thermal contribution, the irradiation was carried out in sequence of 60s each. The alteration of surface morphology or the evolution in nano/micro pattern on the surface is absent during this process. Further, the ion beam irradiation is a synergetic effect i.e., the absence of any measurable change between the sequential and continuous irradiation process. These observations confirm that the formation of dense micro-structures is predominantly governed by fluence-dependent effects rather than by thermal accumulation.

Shown in Fig. 10.4 are SEM micrographs of copper samples in their pristine state and after irradiation with ion beams at normal incidence for five fluence values: 1.3×10^{18} cm^{-2}, 2.7×10^{18} cm^{-2}, 3.2×10^{18} cm^{-2}, 3.9×10^{18} cm^{-2}, and 4.7×10^{18} cm^{-2}. A cross-sectional SEM image corresponding to the highest fluence is also included. In Fig. 10.4a, the untreated Cu surface exhibits small bright features. These arise from the random nanoscale roughness introduced during HNO_3 etching, performed to remove the copper hydroxide layer; this process creates initial nucleation sites that later guide surface evolution. At a fluence of 1.3×10^{18} cm^{-2} (Fig. 10.4b), the surface begins to develop mound-shaped features encircled by distinct rims. This morphology originates from surface destabilization caused by implanted ionic species once their flux exceeds a critical level. As illustrated in Fig. 10.4c, the mound boundaries sharpen further, and early conical protrusions begin to emerge. With continued increase in fluence (Figs. 10.4d–f), the features transform into well-defined micro-hillocks with clear base outlines and increasing height. The surface morphology becomes more uniform, and the conical nature of the structures becomes more evident. At the highest fluence of 4.7×10^{18} cm^{-2} (Fig. 10.4f), densely packed, self-organized micro-cones dominate the surface. These conical elevations typically exhibit heights of 2–4 μm and maintain an approximately constant base diameter of ~ 3 μm. Beyond 5×10^{18} cm^{-2}, the structures begin to lose sharpness, with peak deformation occurring due to enhanced sputtering. At very high ion doses, sputter erosion becomes significant, and the extent of damage depends on both beam intensity and exposure duration [129]. Therefore, appropriate fluence selection is crucial, as optimized fluence not only preserves feature integrity but also increases the density of emitters through the creation of multiple stabilized nucleation centers.

The estimation of surface roughness for both the pristine and ion-irradiated copper samples was carried out using atomic force microscopy (AFM). In particular, samples exposed to ion fluences of 3.2×10^{18} cm^{-2} and 4.7×10^{18} cm^{-2} were examined to quantify the evolution of topographical features. The corresponding variations in work function across these surfaces were evaluated using Kelvin Probe Force Microscopy (KPFM). This AFM-based technique enables nondestructive mapping of local surface potential and work function changes with high spatial resolution. In KPFM, the contact potential difference (V_c) between the probe tip and the surface is measured and is given by:

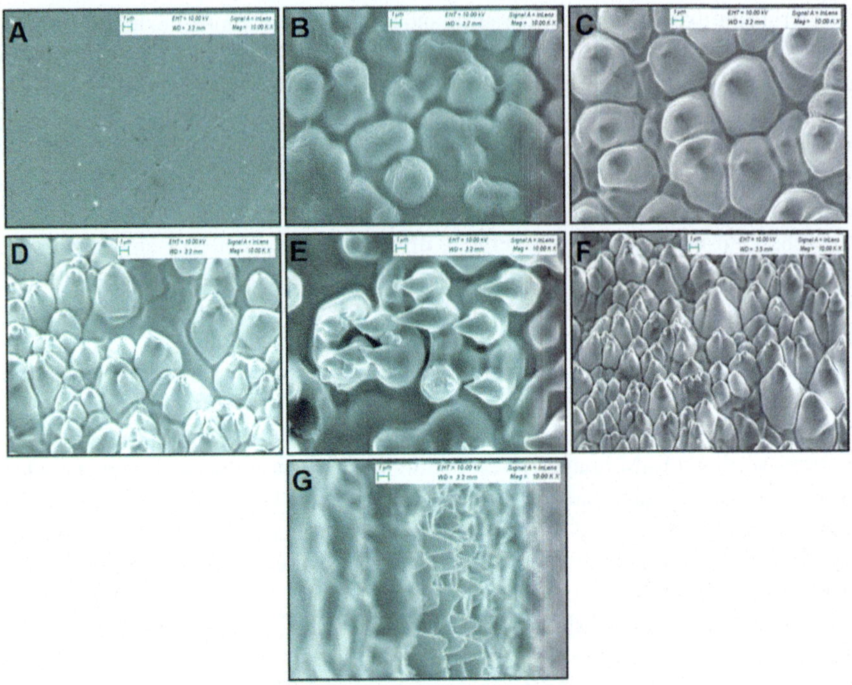

Fig. 10.4 SEM images (10 μm × 10 μm scan area) of **a** pristine Cu sheet, 2 keV Ar + bombarded Cu sheet at normal incidence with fluence; **b** 1.3×10^{18} cm^{-2}, **c** 2.7×10^{18} cm^{-2}, **d** 3.2×10^{18} cm^{-2}, **e** 3.9×10^{18} cm^{-2}, **f** 4.7×10^{18} cm^{-2}, and **g** cross-sectional SEM at 4.7×10^{18} cm^{-2}. Reprinted from [107] Frontiers, under CC-BY license [29]

$$V_c = e\left(\phi_{\text{tip}} - \phi_{\text{sample}}\right), \tag{10.9}$$

where ϕ_{tip} (~ 5.0 eV) and ϕ_{sample} are the work functions of the tip and sample, respectively, while e is the electronic charge. The work function of the irradiated samples is determined by examining the local surface potential variations that arise from topographical undulations and the presence of implanted ionic species. In KPFM measurements, the detected surface potential is effectively averaged over a finite interaction area beneath the probe; as a result, the measured signal reflects contributions from a larger portion of the tip rather than solely its nanometric apex. Consequently, the extracted work function corresponds to an effective value representative of the region scanned by the stylus. Because the ion irradiation is spatially uniform, a net reduction in the average work function is observed. Detailed inspection of the modified surfaces reveals distinct morphological features—elevated micro-conical tips and lower-lying valley regions. In the valleys, the work function is approximately 4.85 eV, whereas at the tip regions it drops significantly to about 4.69 eV for the sample irradiated at a fluence of 4.7×10^{18} cm^{-2}. In contrast, the pristine copper surface exhibits only minor work function variations, ranging from 4.97 to

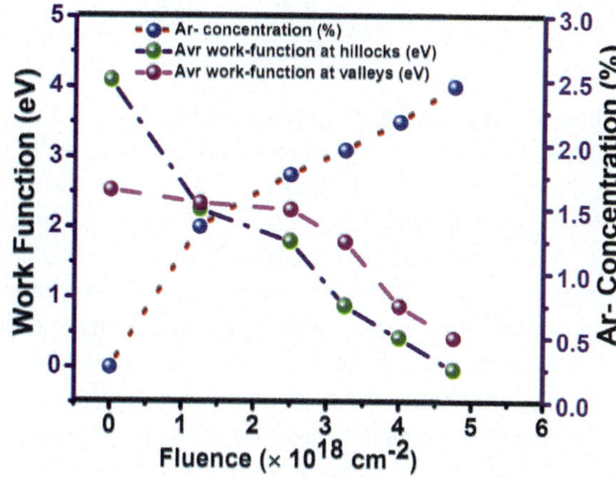

Fig. 10.5 Variation of argon concentration (calculated from XPS data) and variation of the work function distributed over mounds (micro-hill regions) and relatively flat regions (obtained from KPFM data) with ion beam fluences. Avr implies data that are averages at the respective region of hillock or valley

5.09 eV. This narrow distribution arises from the absence of implanted argon ions and minimal height fluctuations (below 100 nm), resulting in a comparatively uniform surface.

The progressive reduction in work function across the hillock (micro-conical tip) regions and the adjacent valley areas as a function of implanted argon concentration is illustrated in Fig. 10.5. Since the implanted ion density increases with longer exposure to the incident ion flux, a systematic decrease in work function is expected and the experimental results follow this trend. Notably, the influence of Ar-ion incorporation on the electronic properties of metallic substrates, particularly on the work function, remains largely unexplored. A recent study has begun addressing this gap by examining Ar^+ irradiation effects in simple metals such as Al and Zn at 30 keV under room-temperature conditions, employing KPFM to provide detailed insight into the underlying theoretical mechanisms of work-function modulation [109]. The experimentally measured average work-function values for both the elevated (hillock) and surrounding (valley) regions, determined using KPFM, are plotted in Fig. 10.5 and clearly exhibit this fluence-dependent behaviour.

10.3.2 Experimental Plotting of the Field Emission Parameters

The ion beam-induced micro-conical structures formed on Cu surfaces were subsequently utilized for field-emission characterization. The electron emission behaviour was analyzed within the framework of Fowler-Nordheim (FN) tunnelling theory. In terms of the experimentally measured emission current (I) and the applied voltage (V), the FN relation is expressed as follows [130]:

$$\ln\left(\frac{I}{V^2}\right) = -\frac{Bd\phi^{3/2}}{V\beta} + \ln\left(\frac{AS\beta^2}{\phi^{3/2}d^2}\right), \tag{10.10}$$

where ϕ is the bulk work function (4.99 eV), d is the gap distance, S is the effective area, and A and B are constants as defined earlier in Eq. 10.8. The current-voltage (I–V) characteristics for the irradiated copper surfaces are presented in Fig. 10.6a, with the corresponding Fowler-Nordheim (FN) representations shown in Fig. 10.6b. The experimental trends clearly indicate a systematic reduction in the effective work function as the ion fluence increases. This decrease correlates with the progressive development of atomic-scale heterogeneity and the emergence of well-aligned, uniformly distributed micro-conical protrusions, as illustrated previously in Fig. 10.4. The combined influence of structural evolution and impurity-induced modifications facilitates more efficient electron extraction under applied electric fields, consistent with the observed FN response. Consequently, samples exposed to higher ion fluence exhibit substantially improved field emission performance, requiring noticeably lower applied voltages to initiate electron emission. As illustrated in Fig. 10.6a, the turn-on voltage drops markedly from 6.51 kV for the pristine copper surface to 2.33 kV for the specimen irradiated at a fluence of 4.7×10^{18} cm^{-2}.The FN plot in Fig. 10.6b is fitted with a straight line and the field enhancement factor (β) is extracted from the slope, given by

$$\beta = -\frac{(6.8 \times 10^3)\phi^{3/2}d}{\text{slope}}. \tag{10.11}$$

The field-enhancement factor is strongly influenced by both the surface geometry and the local work function, as it governs the amplification of the electric field at micro-/nano-scale protrusions. A clear increase in the slope of the linear

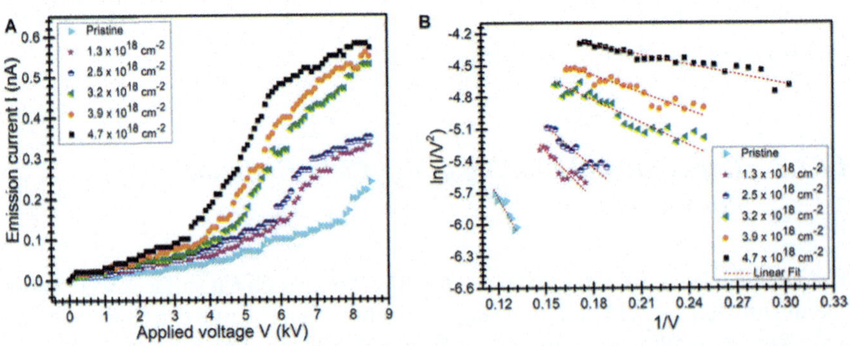

Fig. 10.6 **a** Characteristics of field emission current (I) variation with the applied electric field (V) plot for individual fluences. **b** Fowler-Nordheim equation plot of pristine and irradiated Cu sheets by 2 keV argon ions beam. Voltage is measured in kilovolts, while current is measured in nanoampere. Reprinted from [107] Frontiers, under CC-BY license [29]

FN fits, and therefore in the corresponding enhancement factor is observed as the ion beam fluence is raised. As noted earlier, the standard Fowler-Nordheim framework is derived for an ideal flat metallic surface with a uniform work function. Real irradiated substrates deviate substantially from this assumption. Both the emergence of protruding features, which amplify the local electric field, and the incorporation of subsurface Ar ions modify the effective work function, thereby influencing the emission behavior. Consequently, relying on the bulk work function to analyze field emission from ion-irradiated samples can introduce significant errors. To reliably interpret the emission data, we use a parametric description tailored for atomically heterogeneous surfaces where the work function varies spatially due to height modulation, implanted species, and local structural distortions. In this approach, the effective work function is expressed as a weighted linear combination of the values corresponding to elevated and recessed regions. The weighting coefficients γ and δ represent the fractional surface areas occupied by hillocks and valleys, respectively, and are extracted through quantitative image analysis of KPFM and SEM micrographs. ImageJ software is used to segment and classify the hillock-rich regions in the scanned images. By isolating these segments and evaluating their pixel-based areal fractions, the contributions of distinct surface zones to the overall work function and aspect-ratio distribution are obtained. Although ImageJ is conventionally applied in soft-matter and biological imaging for area-fraction calculations, its pixel-density-based methodology is equally suited for the present surface-topography analysis. The effective work function thus becomes ϕ_{eff}, and is given as follows:

$$\phi_{\text{eff}} = \gamma\phi_1 + \delta\phi_2, \tag{10.12}$$

which is employed in Eq. 10.10 in the place of ϕ, and the β obtained is labeled as β^* which is obtained directly from the slope as discussed earlier. ϕ_I and ϕ_2 are area-averaged work functions of hilly and valley regions, respectively.

The variation in work function for both hillock and valley regions as a function of aspect ratio, is illustrated in Fig. 10.7a. As the fluence increases, the aspect ratio systematically decreases owing to the development of sharper and more distinct hillock-valley features. This morphological evolution is accompanied by a reduction in the local work function across both regions, thereby facilitating electron extraction and resulting in an increased field-enhancement factor. Figure 10.7b represents the trend of the enhancement factor alongside the aspect ratio for different ion beam fluences. Electron emission becomes more favorable from structures with reduced aspect ratio, primarily because their pointed geometry and small radius of curvature intensify the local electric field. Fabricating perfectly uniform arrays of sub-micron conical metallic protrusions, with identical height and base diameter, remains challenging. Therefore, an average aspect ratio was computed for each irradiated sample, defined as the ratio of the base width to the corresponding height of the micro-cones, averaged over a large set of protrusions identified through SEM analysis. The extracted aspect ratios fall within 0.7–1.5, characteristic of low aspect ratio yet sharp microstructures capable of supporting efficient electron tunneling, further aided by local surface charge accumulation in their vicinity. The incorporation of argon ions beneath the surface may additionally modulate the local charge distribution, thereby influencing the enhancement factor. To capture this behavior, a modified enhancement factor (β) was calculated using the effective work function (ϕ_{eff}), and the results are compared in Fig. 10.7 with β derived using the bulk work function (ϕ). The consistently high enhancement factors observed in our measurements can be attributed to the combined influence of sharp geometrical features and atomic-scale modifications induced by ion implantation. At the valley regions, the reduction in work function arises primarily from subsurface argon atoms, which likely alter the local electronic density of states, thereby facilitating electron emission. In contrast, the conical hillock regions experience a dual contribution: (i) modification of the near-surface electronic environment due to implanted argon ions, and (ii) strong geometric field amplification arising from their pointed morphology. These effects lie outside the assumptions of the conventional Fowler-Nordheim formulation, which assumes a planar, chemically homogeneous emitter surface. Given that vacuum electronic behavior is extremely sensitive to micro- and nanoscale surface features, relying solely on the bulk work function to estimate field emission parameters can lead to significant inaccuracies. The parametric model developed here addresses this limitation by incorporating the explicit influence of topographical heterogeneity and localized electronic alterations. This approach highlights the necessity of treating ion-irradiated surfaces as complex, spatially varying emitter systems, where micrometric protrusions and atomic-level defects jointly govern emission efficiency. Such refined modeling therefore provides a more realistic evaluation of the enhancement factor and better reflects the true emission characteristics of modified metallic surfaces.

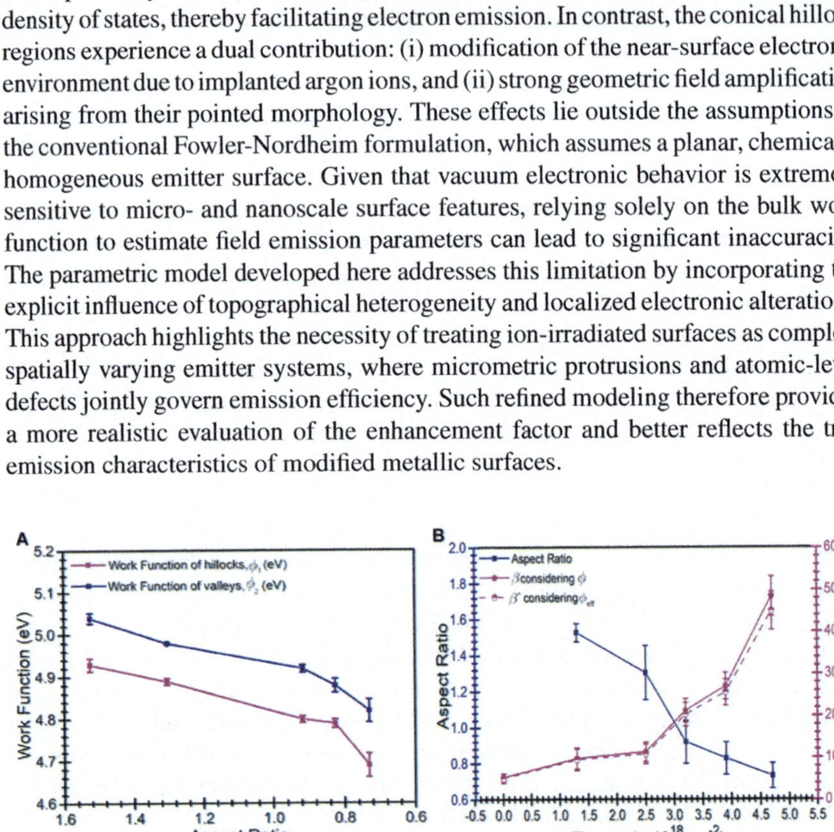

Fig. 10.7 **a** work function of both hillocks and valleys variation with aspect ratio. **b** variation of aspect ratio and field enhancement factor (β and β*) calculated from both bulk (ϕ) and modified work function (ϕ_{eff}) with beam fluence. Reprinted from [107] Frontiers, under CC-BY license [29]

10.3.3 Comparative Study of Field Emission from Micro-conical Tips Created by Ar and Kr Irradiation of Cu Surfaces: Mass Effect

In the present investigation, high-purity pristine Cu foils were irradiated with Ar ions across fluences ranging from (1.3 to 4.7) $\times$ 10^{18} cm^{-2}, and the resulting evolution of micro-conical surface architectures was analyzed, including atomic-scale heterogeneity arising from implanted species and incidental environmental impurities [107]. To further elucidate the mechanisms governing the formation of these protrusions, a comparative investigation was performed using two noble gas ions argon and krypton. Both ion species were delivered at an energy of 2 keV, with the Ar and Kr ion fluxes maintained at 6.47 $\times$ 10^{15} cm^{-2} s^{-1} and 4.81 $\times$ 10^{15} cm^{-2} s^{-1}, respectively. The irradiation duration was adjusted so that the total fluence remained identical for both ions. After generating the micro-conical arrays under equal-fluence conditions, their field-emission characteristics were subsequently assessed. The ability to grow dense, well-defined micro-conical arrays on copper, purely through normal-incidence ion irradiation without the assistance of external catalytic impurities represents a notable experimental advancement [110, 111, 128]. Among the various mechanisms proposed in the earlier sections, the sputter-cone model [113, 114] provides the closest conceptual alignment with the observed morphology. However, unlike the classical description which attributes cone formation to differential sputtering under the presence of a catalyst species with a higher sputter yield than the substrate, the structures produced here emerge in the absence of such foreign impurity catalysts. Instead, the growth appears to be governed by substrate atom redistribution around ion impact sites, driven solely by ion-solid interactions at the applied fluences and energies [132, 132]. The irradiation time directly determines the fluence, such that longer exposures embed a greater number of ions into the copper surface. A digital stopwatch with a resolution of 0.01 s was used to record the irradiation duration, allowing the fluence values to be reported to two decimal places with an estimated uncertainty of approximately 3–5%. As noted earlier, to enable a meaningful comparison between Ar and Kr ion-irradiated samples, the fluence was kept identical for both ion species by appropriately adjusting their respective irradiation times. This ensured that the number of incident ions per unit area remained constant across experiments. To further verify whether thermal effects influenced surface evolution, the irradiation period was divided into repeated 60 s intervals. No measurable differences in surface morphology were observed under this segmented irradiation approach, indicating that the structural formation is not dominated by transient heating effects. A detailed description of the structural evolution from the pristine copper surface to the highest fluence condition under inert-gas ion irradiation has been presented earlier [107].

The formation of well-developed micro-conical features on copper surfaces irradiated with argon ions at a fluence of 3.69 $\times$ 10^{18} cm^{-2} is illustrated in Fig. 10.8a. As the argon fluence is increased to 4.85 $\times$ 10^{18} cm^{-2}, the surface exhibits a more uniform distribution of protrusions with clearly defined conical geometry. At the highest fluence of 5.25 $\times$ 10^{18} cm^{-2}, however, these features begin to degrade due to sputter-induced erosion, as shown in Fig. 10.8b and c. For krypton ion irradiation, pronounced structural development is observed at a fluence of 3.46×10^{18} cm^{-2} (Fig. 10.8d). At a higher fluence of 5.12×10^{18} cm^{-2}, densely packed self-organized nano-cones appear (Fig. 10.8f). Their formation is attributed to substantial mass redistribution, enhanced surface diffusion of adatoms, and implantation of krypton ions. SEM analysis reveals that argon ion-induced cones exhibit heights in the range of 0.06–3 μm with base widths of ~ 2 μm, while the krypton-ion induced structures show heights of 0.04–4 μm and base widths varying between 0.08 and 3 μm. The aspect ratio defined as the ratio of cone base width to height and averaged over multiple protrusions remains relatively low for both cases, consistent with the formation of pointed microstructures.

Figure 10.9 illustrates that the atomic concentration of implanted inert gas ions increases steadily with fluence for both species, consistent with earlier observations. Because krypton has a higher stopping power than argon, attributable to its larger atomic mass, it deposits more efficiently within the copper lattice under identical irradiation conditions. In contrast, for argon irradiation, a slight reduction of roughly 0.56% in implanted ion concentration is observed at the highest fluence. This decrease correlates with the onset of structural erosion at elevated fluences, as evident in Fig. 10.8c. These findings confirm that ion implantation leads to the development of an atomically heterogeneous near-surface region, where the incorporated ions occupy interstitial positions within the substrate.

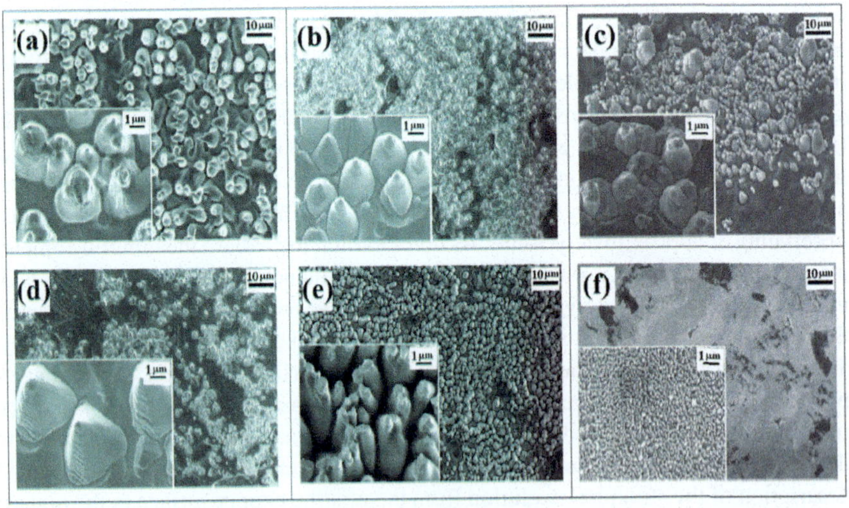

Fig. 10.8 SEM images (10×10 μm^2 scan area for a normal electron beam of incidence, and the tilt angle of substrate is varied within 5° while imaging) of a 2 keV Ar ion bombarded Cu sheet at normal incidence with fluences: **a** 3.69×10^{18} cm^{-2}, **b** 4.85×10^{18} cm^{-2}, and **c** 5.25×10^{18} cm^{-2}, and a Kr ion bombarded Cu sheet at normal incidence with fluences, **d** 3.46×10^{18} cm^{-2}, **e** 4.62×10^{18} cm^{-2}, and **f** 5.12×10^{18} cm^{-2}. Reprinted from [Jayashree Majumdar and Sudeep Bhattacharjee, J. Appl. Phys. 132, 083304 (2022)], with the permission of AIP Publishing

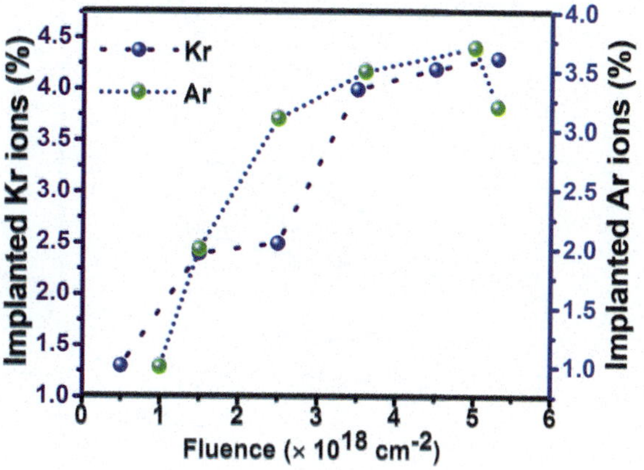

Fig. 10.9 Variation of argon and krypton concentration (calculated from XPS data) with ion beam fluence

Field-emission experiments [133, 134] were conducted on the ion-irradiated copper samples that developed conical nano-micro-scale protrusions on their surfaces. The emission characteristics, plotted as current (*I*) versus applied voltage (*V*), are presented in Fig. 10.10a and b for the argon ion and krypton ion-treated specimens, respectively, while Fig. 10.10c and d provides the corresponding Fowler-Nordheim (FN) representations. For the argon ion-irradiated series, the sample exposed to a fluence of 4.85×10^{18} cm^{-2} exhibited the most favorable performance, showing a minimum turn-on voltage of 2.76 kV and delivering an emission current of 0.58 nA. In contrast, the krypton ion-irradiated sample implanted at 5.12×10^{18} cm^{-2} demonstrated an even lower turn-on voltage of 2.12 kV along with a higher emission current of 0.81 nA. The FN plots were linearly fitted to obtain the geometric field-enhancement factor (β) using the slope of the fitted lines. The work function (ϕ), evaluated through KPFM measurements, decreases for both argon ion and krypton ion modified surfaces as a consequence of irradiation-driven changes in surface morphology and atomic arrangement. The most pronounced improvement is observed for the krypton ion-irradiated sample at 5.12×10^{18} cm^{-2}, which shows significantly enhanced field localization. This enhancement correlates with the presence of more regularly spaced, self-organized surface protrusions, as confirmed by SEM imaging.

The uniformly distributed surface features function as sub-micron or nanoscale emission tips. For pristine copper, the experimentally measured β is 3417, which originates from the nano-grooved texture naturally present on the untreated substrate. These grooves serve as nucleation sites for the growth of well-defined structures once the surface is exposed to ion irradiation at various fluences. After irradiation, the substrates develop arrays of sharp protrusions, which substantially amplify the local electric field under the same applied voltage compared to a flat surface. This geometric intensification greatly increases β, enabling a substantial number of electrons to be emitted. The high localized field assists electrons in surmounting the potential barrier more easily, thereby enhancing field emission. Figure 10.11 presents the variation of β with ion beam fluence for both argon ion and krypton ion-irradiated

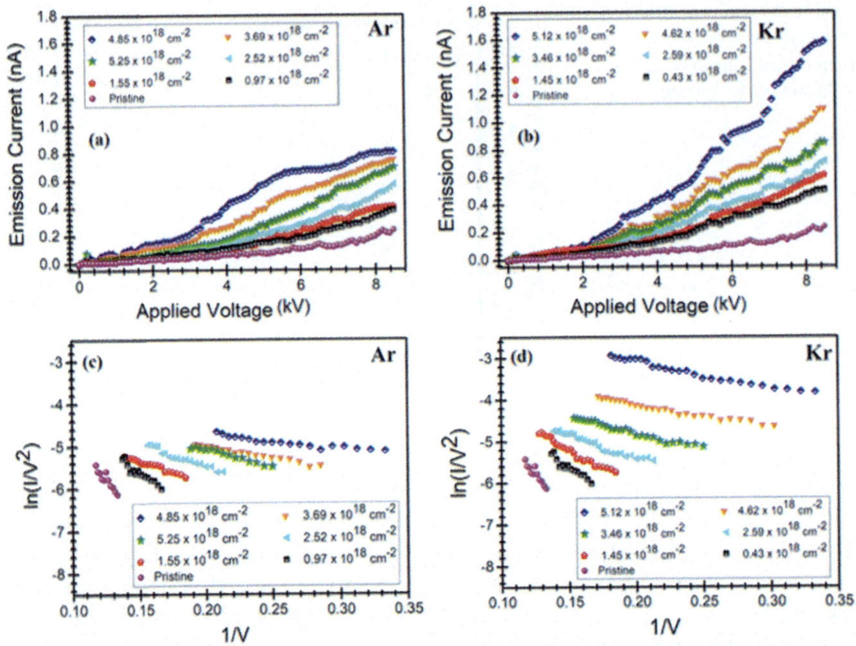

Fig. 10.10 Characteristics of the variation of the field emission current (I) with the applied electric field (V) for individual fluence for **a** argon and **b** krypton ion-irradiated copper surfaces. The Fowler-Nordheim (FN) equation plot of pristine and irradiated Cu samples by a 2 keV argon ion beam in **c** and a 2 keV krypton ion beam in **d**. The voltage is measured in kilo-volts, while current is measured in nanoamperes. Reprinted from [Jayashree Majumdar and Sudeep Bhattacharjee, J. Appl. Phys. 132, 083304 (2022)], with the permission of AIP Publishing

samples. In the case of argon irradiation, β increases steadily up to the maximum fluence of 5.25×10^{18} cm^{-2}. Beyond this point, a sharp decline in β is observed, which is attributed to the considerable erosion and degradation of the surface structures at the highest fluence.

The krypton ion-irradiated copper substrates exhibit progressively higher electron emission with increasing fluence, which is directly linked to the evolution of well-defined surface features characterized by low aspect ratios, pointed tip geometries, and very small radii of curvature. Such morphological elements facilitate efficient electron tunneling because the localized electric fields generated around these protrusions enhance surface charge accumulation. The elevated β values observed for these samples stem from the combined effects of sharp nanoscale irregularities, implantation of inert gas ions, and the incorporation of adatoms on the surface. A precise quantitative breakdown of the contribution from each of these factors will require further controlled experiments in which their individual influences can be isolated. The conical Cu nanostructures fabricated through ion irradiation in this study demonstrate excellent field-emission characteristics, as evidenced by

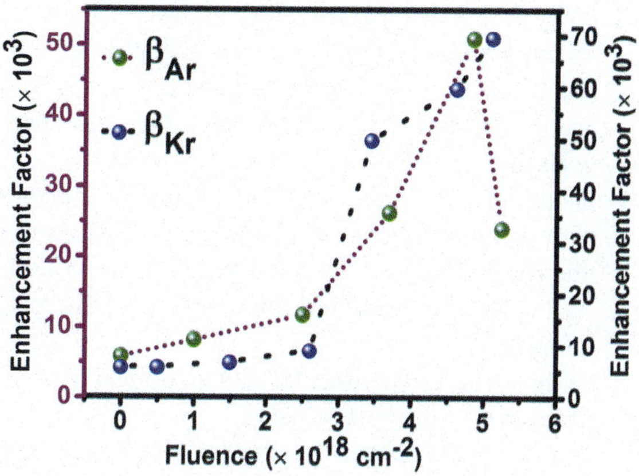

Fig. 10.11 Variation of field enhancement factor (β) of argon and krypton ion-irradiated samples with beam fluence

their high enhancement factors, highlighting their potential for application in next-generation technologies such as flat-panel displays and high-brightness electron emitters. Although krypton and argon ions were irradiated with the same incident energy (2 keV), krypton's higher atomic mass enables it to impart greater momentum during impact. To restore thermodynamic stability, surface atoms redistribute this momentum, fostering enhanced self-organization and leading to a more uniform structural evolution. Consequently, at elevated fluences, krypton ion irradiation suppresses structural dimensions more strongly compared to argon ions. A comparable pattern of mass-dependent surface growth behavior (Ar versus Kr) as a function of ion incidence angle was recently reported by Vandana et al. [135], who observed that implantation with heavier ions (Kr) produces ripples with greater amplitude and reduced wavelength relative to Ar ion irradiation. Moving forward, it would be valuable to investigate in detail how mass redistribution and atomistic rearrangements induced by inert ion implantation drive such self-organized surface evolution. The resulting morphological changes manifest as conical, hillock-like features, and XPS measurements confirm the coexistence of substrate atoms and implanted inert species, validating the formation of an atomically heterogeneous surface.

Questions

Section 10.1

1. Why is it not necessary to measure the reflectivity in infinite range (as predicted by Eq. 10.4) to analyze any particular absorption band for ω_0 in the Kramers-Kronig technique?

2. Why is the Kramers-Kronig analysis not suitable for obtaining the real and imaginary parts of the refractive indices of mixed media (cf. Eq. 10.5 and 10.6)?
3. How are the dielectric constants for the mixed media (ion beam irradiated samples) obtained using the pristine dielectric data procured from the KK technique?
4. How does the energy of incident light and fluence of ion beam change the real and imaginary parts of the dielectric constants? Does the change in fluence of the ion beam affect the interband transition of the material?
5. Which are the two widely employed literature for comparison of optical constants of metals?
6. Why does the experimental data for optical constants of metallic thin films deviate from that of the bulk material?
7. Why does the refractive index (real and imaginary parts) depend upon the wavelength of incident light? What are the different methods by which the refractive index of a material can be obtained?
8. What is meant by skin depth of a metal? On what factors does skin depth depend? Why is the skin depth expected to change for atomically heterogeneous systems? What would happen to the transmission probability of an electromagnetic wave if the skin depth increases?
9. Why is the skin depth largest in case of silver and smallest for aluminum metallic thin films? Calculate the skin depth for Copper and Aluminum, for $\lambda = 405$ nm incoming light and $k = 2.15$ and 4.86 respectively.

Section 10.2

1. Can you describe the effect of charged particles in space such as energetic protons and alpha particles including cosmic rays (protons, electrons and heavy ions), on spacecraft and space vehicles including optical components used in them?
2. What are the typical kinetic energies of alpha particles and protons coming from the solar wind? Can such energies of particles be realized in a laboratory experiment? Ions from plasmas of which gases can be suitable to generate protons and alpha particles?
3. In the experiments the fluence of He ions has been chosen to represent the four (1.1×10^{16} cm^{-2}) and six (1.56×10^{16} cm^{-2}) years doses of alpha particles encountered by a solar orbiter. Can you think of any assumption that is made in calculation of this fluence?
4. Why has it been possible to reduce the irradiation time in our experiments for obtaining the required doses (fluences) corresponding to the 4-year and 6-year journey of a solar orbiter?
5. What could be a possible reason that the reflectivity does not decrease for all energies of ion beams, and tends to again increase at the highest energy (e.g. 3 keV)?

6. What are the principal causes for the change in reflectivity of the metallic thin film samples due to ion irradiation, in addition to the inhomogeneity effect?
7. What is Debye Waller (DW) equation (Eq. 10.7) used for? Considering $\theta = 8°$, calculate the reflectivity from the DW equation for Al at 670 nm and for He ions of 2.0 keV energy and 1.56×10^{16} cm^{-2} fluence, how does the reflectivity compare with the experimental result?
8. Why does helium ion implantation have less effect on optical properties (reflectivity, transmissivity, and absorptivity) of metallic thin films as compared to argon ion implantation?

Section 10.3

1. What are the principal qualities that are desirable in electron field emitters for applications in areas such as field emission cathodes and flat panel displays?
2. What are the factors that modify field emission properties of materials?
3. What are the different ways in which field emission properties of a material can be improved?
4. What are sputter-cones? How can they be created?
5. What is dark current? How can they be prevented in applications such as accelerators, fusion devices, plasma facing components?
6. Why is copper considered quite suitable for field emission studies in metals?
7. What are the different types of self-organized structures in metals and their alloys, created due to ion beam irradiation? What are the typical energies of the ion beams in which these structures are created?
8. In ion beam irradiation of substrates, which parameters of ion beams can be used to control the nature of self-assembly in the host substrate, so that self-assembly of field emitters of desired geometry can be produced cost-effectively and without any external catalytic ions?
9. Please explain the theoretical propositions and the model of the following scientists with regard to ion beam irradiation induced effects on materials: Ehrlich–Schwoebel, Martynenko and Nagel, Bradley–Harper, Robinson and Rossnagel, Davidovitch, and Perkinson.
10. What is field emission? How does field emission occur and on what factors does it depend? Describe and discuss the equation to explain field emission in materials.
11. What effect does the radius of curvature of the field emitters, and the modification of the work function of the material due to formation of microstructures, have on the field emission properties of the materials? Describe the importance of the work of Kyritsakis and Biswas and Djurabekova and Ruzibaev in this regard.
12. Why is differential vacuum pumping necessary in the ion extraction region? Starting their journey in the plasma sheath edge, with what energy does ions hit the substrate meant for ion beam irradiation?

13. What are the typical ion beam fluences used in the experiments? Calculate the number of ions falling on the sample surface of diameter 1 cm when the ion beams fluence is 4.7×10^{18} cm^{-2}? Why are the samples after irradiation kept in vacuum desiccators?
14. Why is optimization of ion beam fluence required for creation of a desired aspect ratio of the micro-conical structures on metallic copper? As one increases the fluence from 1.3×10^{18} cm^{-2} to 4.7×10^{18} cm^{-2}, describe the change in the surface morphology of developing structures on copper substrates due to ion beam irradiation? For fluences higher than 5×10^{18} cm^{-2}, what was observed regarding the sharpness of the generated structures.
15. How is the work function of the sample determined using the KPFM technique? Can you comment on the work function changes of the pristine copper sample, with the copper sample that has implanted argon ions with more or less uniform height distribution?
16. What information does X-ray photoelectron spectroscopy (XPS) spectra of a material provide? What are the typical percentages of embedded Argon inside the Ar ion beam irradiated Cu samples?
17. By what percentage does the average work function of the valley and hillock regions change as the ion beam fluence is increased from unirradiated (pristine) to those irradiated by a fluence of 4.7×10^{18} cm^{-2}?
18. How has the change in work function of the valley and hilly regions of the ion beam irradiated surfaces been accounted for, while undertaking the emission current versus applied voltage plot and the corresponding F-N plot?
19. What happens to the field enhancement factor as a result of the implanted Ar ions inside the substrate? What are the possible reasons that the field enhancement factor increases with increase in ion beam implantation (cf Fig. 10.6b)?
20. Comment on the effect of ion beam fluence on the aspect ratio of the generated micro-conical structures.

Section 10.3.3

1. Comment on the structural changes of the micro-protusions namely change in the aspect ratio (base width/base height) and areal density distribution with increase in fluence for both Ar and Kr ions? What changes are noticed for the density of the micro-conical protusions created on the sample with increase in ion beam fluence?
2. Why are oxygen and carbon seen in the near surface atomic layers of the irradiated samples as observed through XPS (X-ray photoelectron spectroscopy)?
3. In Fig. 10.9, where the deposition of the Ar and Kr ions on Cu substrate is investigated as a function of ion beam fleunce, why does more number of Kr ions gets deposited?
4. Why does the work function of the valley and hillock regions change with increase in ion beam fluence?

5. Explain why the field enhancement factor increases with increase in ion beam fluence for both Ar and Kr? Why do we observe a sudden decrease in the enhancement factor for Ar at ~ 4.8×10^{18} cm^{-2}? There seems to be a sudden increase in the field enhancement factor at ~ 2.6×10^{18} cm^{-2} for both Ar and Kr ions, why is this so?
6. What is sputtering? Comment on the impurities that can be introduced from the plasma and the plasma confining materials due to sputtering?
7. Explain the working principle of a residual gas analyzer. What are the main impurities from the plasma as detected by the residual gas analyzer?

References

1. Ö. Bazkir, Determination of optical constants of silicon photodiode from reflectivity measurements at normal incidence of light. Opt. Lasers Eng. **45**, 245–248 (2007)
2. K.F. Palmer, M.Z. Williams, Determination of the optical constants of a thin film from transmittance measurements of a single film thickness. Appl. Opt. **24**, 1788–1798 (1985)
3. P. Sharma, S.C. Katyal, Determination of optical parameters of a-$(As_2Se_3)90Ge_{10}$ thin film. J. Phys. D. Appl. Phys. **40**, 2115–2120 (2007)
4. R.E. Denton, R.D. Campbell, S.G. Tomlin, The determination of the optical constants of thin films from measurements of reflectance and transmittance at normal incidence. J. Phys. D. Appl. Phys. **5**, 852–863 (1972)
5. V.P. Kutavichus, V.V. Filippov, V.H. Huzouski, Determination of optical parameters and thickness of weakly absorbing thin films from reflectance and transmittance spectra. Appl. Opt. **45**, 4547–4553 (2006)
6. R. Nitsche, T. Fritz, Determination of model-free Kramers-Kronig consistent optical constants of thin absorbing films from just one spectral measurement: application to organic semiconductors. Phys. Rev. B **70**, 195432 (2004)
7. P.B. Johnson, R.W. Christy, Optical constants of the noble metals. Phys. Rev. B **6**, 4370–4379 (1972)
8. S. Babar, J.H. Weaver, Optical constants of Cu, Ag, and Au revisited. Appl. Opt. **54**, 477–481 (2015)
9. H.U. Yang, J. D'Archangel, M.L. Sundheimer, E. Tucker, G.D. Boreman, M.B. Raschke, Optical dielectric function of silver. Phys. Rev. B **91**, 235137 (2015)
10. D. Rioux, S. Vallières, S. Besner, P. Muñoz, E. Mazur, M. Meunier, An analytic model for the dielectric function of Au, Ag, and their alloys. Adv. Opt. Mater. **2**, 176–182 (2014)
11. P.G. Etchegoin, E.C. Le Ru, M. Meyer, An analytic model for the optical properties of gold. J. Chem. Phys. **125**, 164705 (2006)
12. M.A. Ordal, R.J. Bell, R.W. Alexander, L.L. Long, M.R. Querry, Optical properties of fourteen metals in the infrared and far infrared: Al, Co, Cu, Au, Fe, Pb, Mo, Ni, Pd, Pt, Ag, Ti, V, and W. Appl. Opt. **24**, 4493–4499 (1985)
13. E. Palik, *Handbook of Optical Constants of Solids* (Academic Press, 1998), pp. 1–4
14. A.B. Kuzmenko, Kramers-Kronig constrained Variational analysis of optical spectra. Rev Scientific Instrum. **76**, 083108 (2005)
15. P. Grosse, V. Offermann, Analysis of reflectance data using the Kramers Kronig relations. Appl. Phys. A Mater. Sci. Process. **52**, 138–144 (1991)
16. K. Yamamoto, A. Masui, Complex refractive index determination of bulk materials from infrared reflection spectra. Appl. Spectrosc. **49**, 639–644 (1995)
17. G.A. Niklasson, C.G. Granqvist, O. Hunderi, Effective medium models for the optical properties of inhomogeneous materials. Appl. Opt. **20**, 26–30 (1981)

18. A. Sihvola, Mixing rules with complex dielectric coefficients. Subsurface Sensing Tech. Appl. **1**, 393–415 (2000)
19. M.Y. Koledintseva, R.E. DuBroff, R.W. Schwartz, A Maxwell Garnett model for dielectric mixtures containing conducting particles at optical frequencies. Pier **63**, 223–242 (2006)
20. G. Videen, P. Chýlek, Scattering by a composite sphere with an absorbing inclusion and effective medium approximations. Opt. Commun. **158**, 1–6 (1998)
21. K. Pal Singh, J. Majumdar, S. Bhattacharjee, Tuning optical properties of atomically heterogeneous systems created by plasma-based low-energy ion beams. Appl. Opt. **59**, 4507–4516 (2020)
22. K.P. Singh, S. Bhattacharjee, Optical Parameters of Atomically Heterogeneous Systems Created by Plasma Based Low Energy Ion Beams: Wavelength Dependence and Effective Medium Model. Front. Phys. **9**, 671137 (2021)
23. A. Antonaia, M.L. Addonizio, S. Esposito, M. Ferrara, A. Castaldo, A. Guglielmo, et al., Adhesion and structural stability enhancement for Ag layers deposited on steel in selective solar coatings technology. Surf. Coat. Tech. **255**, 96–101 (2014)
24. L. Frey, P. Parrein, L. Virot, C. Pellé, J. Raby, Thin film characterization for modeling and optimization of silver-dielectric color filters. Appl. Opt. **53**, 1663–1673 (2014)
25. J. Ge, Y. Li, J. Wang, Y. Pu, W. Xue, X. Liu, Green synthesis of graphene quantum dots and silver nanoparticles compounds with excellent surface enhanced Raman scattering performance. J. Alloys Compd. **663**, 166–171 (2016)
26. M. Watanabe, K. Kajikawa, An optical fiber biosensor based on anomalous reflection of gold. Sensors Actuators B Chem. **89**, 126–130 (2003)
27. E. Shiles, T. Sasaki, M. Inokuti, D.Y. Smith, Self-consistency and sum-rule tests in the Kramers-Kronig analysis of optical data: applications to aluminum. Phys. Rev. B **22**, 612–628 (1980)
28. H. Ehrenreich, H.R. Philipp, Optical properties of Ag and Cu. Phys. Rev. **128**, 1622–1629 (1962)
29. https://creativecommons.org/licenses/by/4.0/
30. M. Pelizzo, A.J. Corso, P. Zuppella, D. Windt, G. Mattei, P. Nicolosi, Stability of extreme ultraviolet multilayer coatings to low energy proton bombardment. Opt. Express **19**(16), 14838–14844 (2011)
31. J. Barth, C. Dyer, E. Stassinopoulos, Space, atmospheric, and terrestrial radiation environments. IEEE Trans. Nucl. Sci. **50**(3), 466–482 (2003)
32. M.G. Pelizzo, A.J. Corso, E. Tessarolo, R. Botter, R. Hubner, E. Napolitani, M. Bazzan, M. Rancan, A.L.W. Jark, D. Eichert, A. Martucci, Morphological and functional modifications of optical thin films for space applications irradiated with low-energy helium ions. ACS Appl. Mater. Interfaces **10**(40), 34781–34791 (2018)
33. R. Livengood, S. Tan, Y. Greenzweig, J. Notte, S. McVey, Subsurface damage from helium ions as a function of dose, beam energy, and dose rate. J. Vac. Sci. Technol., B: Microelectron. Nanometer Struct.–Process., Meas., Phenom. **27**(6), 3244–3249 (2009)
34. M.G. Pelizzo, A.J. Corso, G. Santi, R. Hübner, D. Garoli, D. Doyle, P. Lubin, A.N. Cohen, J. Erlikhman, G. Favaro, M. Bazzan, J. Drobny, D. Curreli, M. Umansky, Dependence of the damage in optical metal/dielectric coatings on the energy of ions in irradiation experiments for space qualification. Sci. Rep. **11**(1), 3429 (2021)
35. I.D. Sarcina, M.L. Grilli, F. Menchini, A. Piegari, S. Scaglione, A. Sytchkova, D. Zola, Behavior of optical thin-film materials and coatings under proton and gamma irradiation. Appl. Opt. **53**(4), A314–A320 (2014)
36. Q. Wei, H. Liu, D. Wang, S.-X. Liu, Degradation in optical reflectance of Al film mirror induced by proton irradiation. Thin Solid Films **519**(15), 5046–5049 (2011)
37. M. Nardello, P. Zuppella, V. Polito, A.J. Corso, S. Zuccon, M. Pelizzo, Stability of EUV multilayer coatings to low energy alpha particles bombardment. Opt. Express **21**(23), 28334–28343 (2013)
38. A.D. Rousseau, D.L. Windt, B. Winter, L. Harra, H. Lamoureux, F. Eriksson, Stability of EUV multilayers to long-term heating, and to energetic protons and neutrons, for extreme

solar missions, in *Optics for EUV, X-Ray, and Gamma-Ray Astronomy*, ed. by O. Citterio, S.L. O'Dell, vol. II, vol. 5900, (International Society for Optics and Photonics (SPIE, 2005), pp. 14–22
39. E. Antonucci, S. Fineschi, G. Naletto, M. Romoli, D. Spadaro, G. Nicolini, P. Nicolosi, L. Abbo, V. Andretta, A. Bemporad, F. Auchère, A. Berlicki, R. Bruno, G. Capobianco, A. Ciaravella, G. Crescenzio, V.D. Deppo, R. D'Amicis, M. Focardi, F. Frassetto, P. Heinzel, P.L. Lamy, F. Landini, G. Massone, M.A. Malvezzi, J.D. Moses, M. Pancrazzi, M.-G. Pelizzo, L. Poletto, U.H. Schühle, S.K. Solanki, D. Telloni, L. Teriaca, M. Uslenghi, Multi element telescope for imaging and spectroscopy (METIS) coronagraph for the solar orbiter mission, in *Space Telescopes and Instrumentation 2012: Ultraviolet to Gamma Ray*, ed. by T. Takahashi, S.S. Murray, J.-W.A. den Herder, vol. 8443, (International Society for Optics and Photonics (SPIE, 2012), pp. 68–79
40. M.G. Pelizzo, D. Gardiol, P. Nicolosi, A. Patelli, V. Rigato, Design, deposition, and characterization of multilayer coatings for the ultraviolet and visible-light coronagraphic imager. Appl. Opt. **43**(13), 2661–2669 (2004)
41. A.S. Kuznetsov, M.A. Gleeson, F. Bijkerk, Hydrogen-induced blistering mechanisms in thin film coatings. J. Phys. Condens. Matter **24**(5), 052203 (2012)
42. I. Jain, G. Agarwal, Ion beam induced surface and interface engineering. Surf. Sci. Rep. **66**(3–4), 77–172 (2011)
43. S. Chatterjee, K.P. Singh, S. Bhattacharjee, Wetting hysteresis of atomically heterogeneous systems created by low energy inert gas ion irradiation on metal surfaces: Liquid thin film coverage in the receding mode and surface interaction energies. Appl. Surf. Sci. **470**, 773–782 (2019)
44. S. Chatterjee, S. Bhattacharjee, S.K. Maurya, V. Srinivasan, K. Khare, S. Khandekar, Surface wettability of an atomically heterogeneous system and the resulting intermolecular forces. Europhys. Lett. **118**(6), 68006 (2017)
45. A. Chowdhury, S. Bhattacharjee, Experimental investigation of change in sheet resistance and debye temperatures in metallic thin films due to low-energy ion beam irradiation. J. Phys. D. Appl. Phys. **46**(43), 435304 (2013)
46. K.P. Singh, S. Bhattacharjee, Stability of Al and Ag metallic thin film mirrors in a space environment under the implantation of low energy helium ions. Opt. Continuum **1**, 660–673 (2022)
47. S. Zuccon, E. Napolitani, E. Tessarolo, P. Zuppella, A.J. Corso, F. Gerlin, M. Nardello, M.G. Pelizzo, Effects of helium ion bombardment on metallic gold and iridium thin films. Opt. Mater. Express **5**(1), 176–187 (2015) Research Article Vol. 1, No. 4/15 Apr 2022/Optics Continuum 673
48. H. Ehrenreich, H.R. Philipp, Optical properties of Ag and Cu. Phys. Rev. **128**(4), 1622–1629 (1962)
49. A.D. Rakić, Algorithm for the determination of intrinsic optical constants of metal films: application to aluminum. Appl. Opt. **34**(22), 4755–4767 (1995)
50. D. De Boer, Influence of the roughness profile on the specular reflectivity of x rays and neutrons. Phys. Rev. B **49**(9), 5817–5820 (1994)
51. S.T. Lam, Overview of field-emission information storage devices. J. Vac. Sci. Technol. B **21**, 479–482 (2003)
52. N. Egorov, E. Sheshin, *Field Emission Electronics* (Springer, Cham, 2017)
53. A.A. Kuznetzov, S.B. Lee, M. Zhang, R.H. Baughman, A.A. Zakhidov, Electron field emission from transparent multiwalled carbon nanotube sheets for inverted field emission displays. Carbon **48**, 41–46 (2010)
54. F. Jin, A. Beaver, High thermionic emission from barium strontium oxide functionalized carbon nanotubes thin film surface. Appl. Phys. Lett. **110**, 213109 (2017)
55. J.S. Han, T. Bredow, D.E. Davey, A.B. Yu, D.E. Mulcahy, The effect of Al addition on the gas sensing properties of Fe_2O_3-based sensors. Sens. Actuators B Chem. **75**, 18–23 (2001)
56. K.H. Heinig, T. Müller, B. Schmidt, M. Strobel, W. Müller, Interfaces under ion irradiation: Growth and taming of nanostructures. Appl. Phys. A Mater. Sci. Process. **77**, 17–25 (2003)

57. W.L. Chan, E. Chason, Making waves: Kinetic processes controlling surface evolution during low energy ion sputtering. J. Appl. Phys. **101**, 121301 (2007)
58. M. Albrecht, S. Christiansen, J. Michler, W. Dorsch, H.P. Strunk, P.O. Hansson, et al., Surface ripples, crosshatch pattern, and dislocation formation: Cooperating mechanisms in lattice mismatch relaxation. Appl. Phys. Lett. **67**, 232–234 (1995)
59. S. Kajita, T. Yoshida, D. Kitaoka, R. Etoh, M. Yajima, N. Ohno, et al., Helium plasma implantation on metals: Nanostructure formation and visible-light photocatalytic response. J. Appl. Phys. **113**, 134301 (2013)
60. J.-h. Zhu, K. Brunner, G. Abstreiter, Two-dimensional ordering of self assembled Ge Islands on vicinal Si(001) surfaces with regular ripples. Appl. Phys. Lett. **73**, 620–622 (1998)
61. T. Chang, S. Kunuku, K.J. Sankaran, K.-C. Leou, N. Tai, I.-N. Lin, Enhancing the stability of microplasma device utilizing diamond coated carbon nanotubes as cathode materials. Appl. Phys. Lett. **104**, 223106 (2014)
62. J. Liu, B. Zeng, Z. Wu, J. Zhu, X. Liu, Improved field emission property of graphene paper by plasma treatment. Appl. Phys. Lett. **97**, 033109 (2010)
63. C.X. Xu, X.W. Sun, B.J. Chen, Field emission from gallium-doped zinc oxide nanofiber array. Appl. Phys. Lett. **84**, 1540–1542 (2004)
64. A. Gohel, K.C. Chin, Y.W. Zhu, C.H. Sow, A.T.S. Wee, Field emission properties of N_2 and Ar plasma-treated multi-wall carbon nanotubes. Carbon **43**, 2530–2535 (2005)
65. W.-C. Tsai, S.-J. Wang, C.-L. Chang, C.-H. Chen, R.-M. Ko, B.-W. Liou, Improvement of field emission characteristics of tungsten oxide nanowires by hydrogen plasma treatment. Europhys. Lett. **84**, 16001 (2008)
66. I. Tanyeli, L. Marot, M.C.M. van de Sanden, G. De Temmerman, Nanostructuring of iron surfaces by low-energy helium ions. ACS Appl. Mater. Interfaces **6**, 3462–3468 (2014)
67. K. Zhang, O. Bobes, H. Hofsäss, Designing self-organized nanopatterns on Si by ion irradiation and metal co-deposition. Nanotechnology **25**, 085301 (2014)
68. T.C. Kim, C.-M. Ghim, H.J. Kim, D.H. Kim, D.Y. Noh, N.D. Kim, et al., Kinetic roughening of ion-sputtered Pd(001) surface: Beyond the Kuramoto-Sivashinsky model. Phys. Rev. Lett. **92**, 246104 (2004)
69. G. Ozaydin, A.S. Özcan, Y. Wang, K.F. Ludwig, H. Zhou, R.L. Headrick, et al., Real time x-ray studies of Mo-seeded Si nanodot formation during ion bombardment. Appl. Phys. Lett. **87**, 163104 (2005)
70. G.M. Mc Cracken, A review of experimental evidence for arcing and sputtering in tokamaks. J. Nucl. Mater. **93–94**, 3–16 (1980)
71. J.B. You, X.W. Zhang, P.F. Cai, J.J. Dong, Y. Gao, Z.G. Yin, et al., Enhancement of field emission of the Zno film by the reduced work function and the increased conductivity via hydrogen plasma treatment. Appl. Phys. Lett. **94**, 262105 (2009)
72. Z. Zheng, L. Liao, B. Yan, J.X. Zhang, H. Gong, Z.X. Shen, et al., Enhanced field emission from argon plasma-treated ultra-sharp α-Fe_2O_3 nanoflakes. Nanoscale Res. Lett. **4**, 1115–1119 (2009)
73. A. Chowdhury, S. Bhattacharjee, Localized subsurface modification of materials using micro-low-energy multiple ion beamlets. AIP Adv. **1**, 042150 (2011)
74. M. Ritter, M. Stindtmann, M. Farle, K. Baberschke, Nanostructuring of the Cu(001) surface by ion bombardment: a STM study. Surf. Sci. **348**, 243–252 (1996)
75. L. Feng, H. Yan, H. Li, R. Zhang, Z. Li, R. Chi, et al., Excellent field emission properties of vertically oriented Cuo nanowire films. AIP Adv. **8**, 045109 (2018)
76. G.A. Evangelakis, N.I. Papanicolaou, Adatom self-diffusion processes on (001) copper surface by molecular dynamics. Surf. Sci. **347**, 376–386 (1996)
77. Y.W. Zhu, A.M. Moo, T. Yu, X.J. Xu, X.Y. Gao, Y.J. Liu, et al., Enhanced field emission from O_2 and CF_4 plasma-treated Cuo nanowires. Chem. Phys. Lett. **419**, 458–463 (2006)
78. C. Schäfer, D.A. Gollmer, A. Horrer, J. Fulmes, A. Weber-Bargioni, S. Cabrini, et al., A single particle plasmon resonance study of 3D conical Nanoantennas. Nanoscale **5**, 7861–7866 (2013)

79. I. Tanyeli, L. Marot, M.C. van de Sanden, G. De Temmerman, Nanostructuring of iron surfaces by low-energy helium ions. ACS Appl. Mater. Interfaces **6**, 3462–3468 (2014)
80. R. Robinson, S. Rossnagel, Ion-beam-induced topography and surface diffusion. J. Vac. Sci. Technol. **21**, 790–797 (1982)
81. M. Navez, D. Chaperot, C. Sella, Microscopie electronique-etude de lattaque du verre par bombardement ionique. C. R. Hebd. Seances Acad. Sci. **254**, 240 (1962)
82. G. Ozaydin, A.S. Özcan, Y. Wang, K.F. Ludwig, H. Zhou, R.L. Headrick, D.P. Siddons, Real-time x-ray studies of Mo-seeded Si nanodot formation during ion bombardment. Appl. Phys. Lett. **87**, 163104 (2005)
83. A. Keller, S. Facsko, Ion-induced nanoscale ripple patterns on Si surfaces: theory and experiment. Materials **3**, 4811–4841 (2010)
84. R. Pétri, P. Brault, O. Vatel, D. Henry, E. André, P. Dumas, F. Salvan, Silicon roughness induced by plasma etching. J. Appl. Phys. **75**, 7498–7506 (1994)
85. Q. Shi, S. Kajita, N. Ohno, M. Tokitani, D. Nagata, S. Feng, The influence of impurities on the formation of nanocone structures on silicon surface irradiated by low energy helium plasma. J. Appl. Phys. **128**, 023301 (2020)
86. S. Kajita, T. Yoshida, D. Kitaoka, R. Etoh, M. Yajima, N. Ohno, H. Yoshida, N. Yoshida, Y. Terao, Helium plasma implantation on metals: nanostructure formation and visible-light photocatalytic response. J. Appl. Phys. **113**, 134301 (2013)
87. M. Thompson, D. Drummond, J. Sullivan, R. Elliman, P. Kluth, N. Kirby, D. Riley, C. Corr, Effect of W self-implantation and he plasma exposure on early-stage defect and bubble formation in tungsten. Nucl. Fusion **58**, 066010 (2018)
88. D. Hwangbo, S. Kajita, N. Ohno, P. McCarthy, J.W. Bradley, H. Tanaka, Growth of nano-tendril bundles on tungsten with impurity-rich he plasmas. Nucl. Fusion **58**, 096022 (2018)
89. M. Thompson, L. Magyar, C. Corr, Nanoscale modification of silicon and germanium surfaces exposed to low-energy helium plasma. Sci. Rep. **9**, 10099 (2019)
90. S. Kajita, G.D. Temmerman, T. Morgan, S. van Eden, T. de Kruif, N. Ohno, Thermal response of nanostructured tungsten. Nucl. Fusion **54**, 033005 (2014)
91. Y.-W. You, D. Li, X.-S. Kong, X. Wu, C. Liu, Q. Fang, B. Pan, J. Chen, G.-N. Luo, Clustering of H and He, and their effects on vacancy evolution in tungsten in a fusion environment. Nucl. Fusion **54**, 103007 (2014)
92. Y.-P. Xu, H.-S. Zhou, Y.M. Peng, T. Lu, X.-C. Li, F. Liu, J. Wang, H.-D. Liu, Y.-M. Lyu, G.-N. Luo, Surface modification of F82H steel exposed to low energy, high flux He plasmas. Nucl. Fusion **57**, 056038 (2017)
93. T. Chang, S. Kunuku, K.J. Sankaran, K.-C. Leou, N. Tai, I.-N. Lin, Enhancing the stability of microplasma device utilizing diamond coated carbon nanotubes as cathode materials. Appl. Phys. Lett. **104**, 223106 (2014)
94. J. Liu, B. Zeng, Z. Wu, J. Zhu, X. Liu, Improved field emission property of graphene paper by plasma treatment. Appl. Phys. Lett. **97**, 033109 (2010)
95. A.A. Kuznetzov, S.B. Lee, M. Zhang, R.H. Baughman, A.A. Zakhidov, Electron field emission from transparent multiwalled carbon nanotube sheets for inverted field emission displays. Carbon **48**, 41–46 (2010)
96. R.M. Bradley, J.M. Harper, Theory of ripple topography induced by ion bombardment. J. Vac. Sci. Technol. A **6**, 2390–2395 (1988)
97. R. Cuerno, A.-L. Barabási, Dynamic scaling of ion-sputtered surfaces. Phys. Rev. Lett. **74**, 4746 (1995)
98. Y.V. Martynenko, M.Y. Nagel, Model of fuzz formation on a tungsten surface. Plasma Phys. Rep. **38**, 996–999 (2012)
99. S.-T. Lam, Overview of field-emission information storage devices. J. Vac. Sci. Technol. B **21**, 479–482 (2003)
100. F. Jin, A. Beaver, High thermionic emission from barium strontium oxide functionalized carbon nanotubes thin film surface. Appl. Phys. Lett. **110**, 213109 (2017)
101. W.-C. Tsai, S.-J. Wang, C.-L. Chang, C.-H. Chen, R.-M. Ko, B.-W. Liou, Improvement of field emission characteristics of tungsten oxide nanowires by hydrogen plasma treatment. Europhys. Lett. **84**, 16001 (2008)

102. J. You, X. Zhang, P. Cai, J. Dong, Y. Gao, Z. Yin, N. Chen, R. Wang, H. Yan, Enhancement of field emission of the ZnO film by the reduced work function and the increased conductivity via hydrogen plasma treatment. Appl. Phys. Lett. **94**, 262105 (2009)
103. Z. Zheng, L. Liao, B. Yan, J. Zhang, H. Gong, Z. Shen, T. Yu, Enhanced field emission from argon plasma-treated ultra-sharp α Fe_2O_3 nanoflakes. Nanoscale Res. Lett. **4**, 1115–1119 (2009)
104. S. Kajita, N. Ohno, Y. Hirahata, M. Hiramatsu, Field emission property of nanostructured tungsten formed by helium plasma irradiation. Fusion Eng. Des. **88**, 2842–2847 (2013)
105. M. Kildemo, S. Calatroni, M. Taborelli, Breakdown and field emission conditioning of Cu, Mo, and W. Phys. Rev. Spec. Top. Accel. Beams **7**, 092003 (2004)
106. A. Chowdhury, S. Chatterjee, A. Dutta, S. Bhattacharjee, Stopping potential and ion beamlet control for micro-resistive patterning through sub-Debye length plasma apertures. AIP Adv. **4**, 127127 (2014)
107. J. Majumdar, S. Bhattacharjee, Microconical structure formation and field emission from atomically heterogeneous surfaces created by microwave plasma based low-energy ion beams. Front. Phys. **9**, 310 (2021)
108. F. Djurabekova, A. Ruzibaev, E. Holmström, S. Parviainen, M. Hakala, Local changes of work function near rough features on Cu surfaces operated under high external electric field. J. Appl. Phys. **114**, 243302 (2013)
109. A. Horváth, N. Nagy, G. Vértesy, R. Schiller, Phenomenology of the effect of ion irradiation on the work function of metals. Nucl. Instrum. Methods Phys. Res., Sect. B **466**, 12–16 (2020)
110. P. McCarthy, D. Hwangbo, M. Bilton, S. Kajita, J.W. Bradley, Enhanced fuzzy tungsten growth in the presence of tungsten deposition. Nucl. Fusion **60**, 026012 (2020)
111. K. Zhang, O. Bobes, H. Hofsäss, Designing self-organized nanopatterns on Si by ion irradiation and metal co-deposition. Nanotechnology **25**, 085301 (2014)
112. S. Kajita, S. Kawaguchi, N. Yoshida, N. Ohno, H. Tanaka, Morphologies of co depositing W layer formed during He plasma irradiation. Nucl. Fusion **58**, 106002 (2018)
113. G. Wehner, D. Hajicek, Cone formation on metal targets during sputtering. J. Appl. Phys. **42**, 1145–1149 (1971)
114. B. Davidovitch, M.J. Aziz, M.P. Brenner, On the stabilization of ion sputtered surfaces. Phys. Rev. B **76**, 205420 (2007)
115. R.M. Bradley, Theory of nanodot and sputter cone arrays produced by ion sputtering with concurrent deposition of impurities. Phys. Rev. B **83**, 195410 (2011)
116. M. Thompson, K. Song, P. Kluth, N. Kirby, C. Corr, Helium plasma induced nanostructure formation in copper and nickel. Surf. Topogr. Metrol. Prop. **7**, 015007 (2019)
117. P. Nayak, S. Ramaprabhu, Directed self assembly of copper-based hierarchical nanostructures on nitrogen-doped graphene and their field emission studies. J. Phys. Chem. C **119**, 2917–2924 (2015)
118. Y. Zhu, A. Moo, T. Yu, X. Xu, X. Gao, Y. Liu, C. Lim, Z. Shen, C. Ong, A. Wee, et al., Enhanced field emission from O_2 and CF_4 plasma-treated CuO nanowires. Chem. Phys. Lett. **419**, 458–463 (2006)
119. J. Camarero, V. Cros, M.J. Capitán, J. Álvarez, S. Ferrer, M.A. Niño, et al., Epitaxial growth of metals with high Ehrlich-Schwoebel barriers and the effect of surfactants. Appl. Phys. A Mater. Sci. Process. **69**, 553–557 (1999)
120. G.K. Wehner, Cone formation as a result of whisker growth on ion bombarded metal surfaces. J. Vacuum Sci. Technol. A Vacuum Surf. Films **3**, 1821–1835 (1985)
121. J.C. Perkinson, E. Anzenberg, M.J. Aziz, K.F. Ludwig Jr., Model-independent test of the truncated crater function theory of surface morphology evolution during ion bombardment. Phys. Rev. B **89**, 115433 (2014)
122. R.H. Fowler, L. Nordheim, Electron emission in intense electric fields. Proc. R Soc. Lond. Ser. A, Containing Pap. Math. Phys. Character **119**, 173–181 (1928)
123. S. Kajita, N. Ohno, Y. Hirahata, M. Hiramatsu, Field emission property of nanostructured tungsten formed by helium plasma irradiation. Fusion Eng. Des. **88**, 2842–2847 (2013)

124. A. Kyritsakis, J.P. Xanthakis, Derivation of a generalized Fowler-Nordheim equation for nanoscopic field-emitters. Proc. R. Soc. A **471**, 20140811 (2015)
125. D. Biswas, R. Ramachandran, Curvature correction to the field emission current. J. Vacuum Sci. Technol. B **37**, 021801 (2019)
126. H. Toijala, K. Eimre, A. Kyritsakis, V. Zadin, F. Djurabekova, Ab initio calculation of field emission from metal surfaces with atomic-scale defects. Phys. Rev. B **100**, 165421 (2019)
127. S.K. Maurya, S. Barman, S. Paul, S. Bhattacharjee, Charge dissipation and self focusing limit in high current density ion beam transport through a micro glass capillary. J. Phys. D. Appl. Phys. **52**, 055205 (2018)
128. A. Chowdhury, S. Bhattacharjee, Micron-focused ion beamlets. J. Appl. Phys. **107**, 093307 (2010)
129. F. Frost, R. Fechner, B. Ziberi, J. Völlner, D. Flamm, A. Schindler, Large area smoothing of surfaces by ion bombardment: fundamentals and applications. J. Phys. Condensed Matter **21**, 224026 (2009)
130. S. Bhattacharjee, A. Vartak, V. Mukherjee, Experimental study of space charge-limited flows in a Nanogap. Appl. Phys. Lett. **92**, 191503 (2008)
131. S. Kajita, S. Kawaguchi, N. Yoshida, N. Ohno, H. Tanaka, Morphologies of co-depositing W layer formed during He plasma irradiation. Nucl. Fusion **58**, 106002 (2018)
132. R.M. Bradley, Theory of nanodot and sputter cone arrays produced by ion sputtering with concurrent deposition of impurities. Phys. Rev. B **83**, 195410 (2011)
133. R.H. Fowler, L. Nordheim, Electron emission in intense electric fields. Proc. R. Soc. London, Ser. A **119**, 173–181 (1928)
134. S. Bhattacharjee, A. Vartak, V. Mukherjee, Experimental study of space-charge-limited flows in a nanogap. Appl. Phys. Lett. **92**, 191503 (2008)
135. Vandana, P. Chhokkar, S. Kumar, V. Singh, R.K. Pandey, T. Kumar, S. Kumar, et al., Projectile's mass-dependent nanopatterning of Si (1 0 0) for different incidence angles. Mater. Lett. **309**, 131348 (2022)

Chapter 11
Summary and Future Prospects

The book is on the subject of deriving ion beams from plasmas that are created in multicusp magnetic fields. The ion beams are accelerated to low energies in the range 0–3 keV by employing electrostatic fields and are then irradiated onto a variety of materials for modification of their properties. The materials employed in the research are primarily in the form of thin films having thickness of a few hundred nanometers. Plasmas of a variety of gaseous elements have been created including atomic gases such as Ar, He, and Kr and molecular gases such as H_2, and N_2. However, most of the works have been undertaken with ions obtained from inert gas plasmas. The extracted ions having a low energy in the range 0–3 keV are made to impinge on the thin films of metals, and they enter into the surface and subsurface layers of the host material. The penetration depth of these ions lies from a few to tens of nanometers inside the host material. Since the gaseous ions are inert, they do not react with the host material, rather they enter into the lattice of the host, and no chemical compounds are created inside the host. Therefore, introduction of the inert foreign atoms into the host creates an atomically heterogeneous system; however, the chemical identity of both the host and the foreign atoms is retained. A variety of host metallic thin films are employed as substrates in the work, such as Al, Cu, and Ag. Therefore, with multiple choices for the type of projectile ions and the host substrates, a variety of atomically heterogeneous systems can be created and their novel properties such as surface, optical, electrical, and field emission can be investigated. It may be noted that these inert gas ions vary in their atomic mass and therefore transfer different amounts of momentum onto the substrate, which can change the ways ion beam interacts with the substrate, particularly the momentum loss inside the substrate, dislocations of the host atoms, and creation of defects. While it is difficult to measure or observe the phenomena inside the host material during interaction of ion beam with the host, SRIM/TRIM and TRYDYN HDZ simulations helped to understand the ion trajectories, distributions, and ion ranges inside the host. The embedded ionic impurities in the host modify the physical properties of the parent

S. Bhattacharjee, *Atomically Heterogeneous Systems*,
https://doi.org/10.1007/978-981-95-9733-8_11

material. For example, it was found that the electrical sheet resistance of the materials could be tuned to a large extent utilizing the ion beams as discussed toward the end of Chap. 7. The sheet resistance and residual sheet resistance were measured for a variety of ion-irradiated materials from room to low temperatures and discussed in great details in Chap. 8. As a prelude to these discussions, a theoretical background of the origin of resistivity in metals is provided and the electron transport is discussed in single and polycrystalline films, including the temperature dependence of resistivity in a metal. In Chap. 8, the Debye temperatures θ_D are determined from Bloch-Gruneisen fitting of the temperature variation of sheet resistance data and it is found that θ_D decreases with increase in both fluence and ionic mass number, arising primarily from the change in bulk modulus of the nano-film. Chap. 9 and parts of Chap. 10 dwell on the optical properties of the ion beam-modified metallic thin films. The optical constants such as the real and imaginary parts of the refractive index, skin depth, optical conductivity, including basic optical parameters such as reflectivity, transmissivity, and absorptivity have been determined, as a function of several beam parameters such as fluence, energy, and ionic species. The reflectivity, transmissivity, and absorptivity have been measured as a function of wavelength of incident light using a spectrophotometer. The real and imaginary parts of the refractive index were measured using the pseudo-Brewster angle technique. Several other theoretical models such as the Kramers-Kronig method have been applied to determine the optical constants, and the results have been compared with those obtained from the existing mixed medium models such as Maxwell-Garnet and Bruggeman approximations. From the experimental data it was noted that the interband transitions which are characteristic of each metal substrate can affect the optical properties of the ion beam-modified thin films. Part of Chap. 10 provides results of an interesting and important application of ion beam irradiation on metallic thin-film mirrors, which are widely used in spacecrafts and other space vehicles, where the effect of He ion beams with fluences comparable to the 4-year and 6-year cycle of a solar orbiter was investigated. In this chapter, the electron field emission properties of ion beam-modified materials are discussed. The host material properties also influence the nature of the ion-matter interactions. For example, the softness or hardness of the material can decide the degree of sputtering (or atomic removal) when the ion beams impinge upon the material and has a bearing on the rate of redeposition on the substrate. As was seen in Chap. 10, it can change the structure of the micro-conical growths that are created on the surface of the material. It was found that both atomic heterogeneity and aspect ratio of the created structures can affect the field emission from these structures. The basics of field emission are discussed including the underlying theories of micro-structural growths on the surface of metals. Field emission is influenced both by microstructures that are formed, due to enhancement of the electric field, depending upon the aspect ratio of the structures, and due to formation of atomic heterogeneity, i.e., the implanted subsurface atoms, due to lowering of the work function (as was observed in the case of both hillocks and valley regions of the generated structures). X-ray photoelectron spectroscopy and Kelvin Probe Force Microscopy (KPFM) provided important information on the extent of atomic heterogeneity and work function of the ion beam-modified surfaces. In Chap. 10, a comparative field emission study from micro-conical tips created in Cu substrate by two different low-energy noble gas ion beams, namely, Ar and Kr is investigated. The development of the structures (aspect ratio, areal distribution of hilly and valley regions) has been investigated as a function of ion beam fluence and the amount of foreign atomic impurities for each fluence (implanted Kr and Ar atoms). Dynamic scaling theory has been applied to understand the nature of the growth of surface morphology. The structural growths have also been looked at using KPFM and Transmission Electron Microscopy (TEM). The cross-sectional TEM results unveiled the dimensions of the surface growths, the formation of an amorphous and a damage layer.

We now discuss about the future prospects of the works reported in this book. In the present stage, the focused multi-ion beamlets can generate ion beamlets of several gaseous ions such as Ne, Ar, Kr, H_2, and N_2. The ion beamlets are focused to ~ 28 μm spot with beam current density of ~ 0.51 A/cm^2 for Ar ions. By slightly changing the plasma electrodes, we can also generate low-energy broad ion beams (diameter ~ 2 cm). There is much scope of further research in optimizing the MIBS system for experiments. The beamlets can be further focused to sub-micron beam diameter for more effective localized application. For that the aperture diameter of the plasma electrode (PE) needs to be reduced to a few microns (< 5 μm). This is possible with laser micro-drill on very thin metal foils and these foils can be used as PE. Beam current (which is quite high presently) can be reduced for better focusing so that Coulomb repulsion is reduced. Furthermore, the whole pattern can be demagnified on the substrate using suitable beam optics. With individual control on ion beamlets, the MIBS can be used as a pattern generator and the pattern can be directly put on the substrate as direct ion beam lithography where stencil masks are no longer required. This has a huge potential to be commercialized.

In the work presented in this book, low-energy ion beams with energies ranging from 0.1 to 5 keV can be generated with flux of the order ~ 10^{15} ions/cm^2 s. We have primarily seen the irradiation effects on metallic films only. One can look into the irradiation effects in soft materials like polymers [1–3], carbon nanotubes or in graphene. Ion beam easily provides the energy to break the bonds in polymers. Therefore, ion beam implantation can be used to change the surface morphology, color, and structure of the polymer samples. Graphene is one of the most attractive materials for next-generation electronics because of its linear energy momentum dispersion near the Fermi level which leads to its extremely high carrier mobility. Ion beam (preferably low energy) and electron beam are two very important tools for modification of the properties of graphene. Electronic and structural properties of graphene can be changed with low-energy beam irradiation by introducing defects in it [4, 5]. The defects and charge impurities act as scattering centers and limit its electron mobility [6]. Ion beam-induced defects can also cause a significant change in the optical properties of graphene [7] observed from the Raman intensity of the D peak. There are reports on ripple formation and electron delocalization upon 1 keV Ar^+ ion irradiation [8]. Another important property of graphene is the zero band gap at the Fermi level and there are efforts to tune the band gap in a controlled manner by using ion beam [9]. Some theoretical works also predict such changes using boron and nitrogen beam irradiation [10]. Low- and medium-energy beam irradiation of various species (Ar, Kr, N, etc.) can bring about changes in electrical mobility, surface morphology, and band gap in a controlled manner.

Optical and magnetic properties of materials have been tailored by low-energy ion irradiation [11, 12]. We have primarily looked at the modification in the optical properties of metallic thin films. Irradiation-induced defects change the properties and, in some cases, a correlation between the optical and magnetic properties is established [11]. Recently, many experiments have focused on the properties of ZnO nanorods because magnetic ion-doped ZnO nanorods [13–15] exhibit pronounced optical properties, controllable magnetic properties with a Curie temperature. UV-VIS absorption spectroscopy measurements reveal that the doping of Mn^{2+} leads to a red shift of the absorption edge of ZnO nanorods [16]. Another emerging area of low-energy ion beam irradiation is the enhancement of field emission properties by ions in materials like carbon nanotubes (CNT), α-Fe_2O_3, nanoflakes, carbon and ZnO nanotips, etc. Although we have covered this topic of research in metals but they can also be investigated in these materials. Earlier these materials are plasma treated for the enhancement of field emission [17–19]. Low-energy ion beams can be used for this and one can have more control over the process with different ionic species, beam energy, and fluence [20, 21]. Acuna et al. have shown that O^+ ion beam irradiation of nanotubes preferentially eliminates carbon atoms bonded to highly reactive parts of the material by forming volatile carbon oxide compounds. Several such interesting research areas can be undertaken.

In the chapters on investigation of optical properties of ion beam-irradiated materials, the optical parameters of various metallic thin films under the implantation of low-energy gaseous ion beams have been carried out. The implantation of ion beams in these studies has been done at a normal incidence angle. Research has been conducted in the case of low-energy ion beams (0.5 keV for argon ions and 0.5–3 keV for He and H ions). However, the research can be extended to the off-normal incidence angle for low- as well as high-energy ion beam implantation [22]. The off-normal ion beam implantation can create various unique structures (such as ripples, cone like, and faceted structures [22–24]) on the implanted surface. One can investigate the optical properties with varying angles of ion beam irradiation and the tunability of optical properties can be compared with that of normal incidence angle. The electromagnetic wave interacts in a different manner with the structured surface than that of the smooth surface because of the scattering effect. Hence, the optical properties of the metallic thin films for varying incidence angles can also be investigated. Apart from this, one can also irradiate the metallic thin films with more than one ionic species at a time. We have investigated the irradiation effect of a single ionic species on the optical properties of metallic thin films, and it has been found that the tuning of optical properties depends upon the polarizability or dielectric constants of the ionic species being used. Hence, the effect of a mixed ion beam (for example, Ar + Kr, Ar + H, etc.) can be looked at. In the present research, metallic thin films have been utilized for the investigation of optical properties. However, this can be extended to other materials as well such as dielectric, semiconductors, and polymers [25–29]. As the optical properties of the above-mentioned materials have drawn the interest of the scientific community because of their potential applications in photonic devices, bio and optical sensors, including solar cells. The alteration of the optical properties can be achieved in various ways including physical texturing and changing the temperature of the materials. The optical properties of polymers are mainly defined by their transmissivity, refractive index, and dispersion relation [30]. One can tune the refractive index and in turn the optical properties of polymers by irradiating them with low-energy noble gaseous ion beams. The refractive index of the mixed medium is decided by the refractive indices of both host and implanted ions and the amount of the implanted impurities [31]. The effect of irradiation on polymers is expected to be different as compared to the case of metals. Since the optical properties of metals are mainly decided by the behavior of free electrons present in the metals. Hence, the refractive index of polymers can be tuned in a controlled manner by the implantation technique, as we can select the irradiation parameters such as types of impurity ions, their energy, and their dose (or fluence). One can also look at the optical properties of polymers by varying the angle of incidence that can create the structures on the surface. Additionally, the study can be conducted with high-energy noble as well as reactive ion species. One may also examine the effect of low-energy ion beams on the surface plasmon generation and their propagation along the interface. This has a lot of interest in the community. Surface plasmons are electron-density oscillations and are generated on the interface of metal-dielectric [32, 33]. Gold and silver are considered to be the most favorable materials for surface plasmon study, which is important because of their wide range of applications including sensors, bio-photonics, photonic devices, and in subwavelength optics [34–37]. Both the propagation vector and propagation length of surface plasmons depend upon the dielectric constants of both media. The propagation length of surface plasmons sets the upper size limit of the photonic band gap circuit, while the decay length decides the maximum height of the fabricated photonic components and their ratio fixes the maximum number of surface plasmon integrable circuits [34]. Hence, one can tune the dielectric constants of the metallic media using low-energy ion beams and can look at the effects of irradiation. Thus, the controlled tuning of dielectric constants of metallic surfaces gives us the flexibility of altering the propagation length, decay length, and their ratio in a suitable manner, which in turn would be helpful in the fabrication of desired photonic devices. As we have very less choices of materials for surface plasmonic study, this method of irradiation can provide us additional flexibility in tuning the propagation length and wave vector of surface plasmons. The present irradiation technique can provide us various ionic species including noble and reactive species. One can easily change the impinging energy and fluence of incoming ion beams to tune the properties of materials in a controllable manner. This will generate the structures on the metallic surface and hence the properties of surface plasmon (interaction of surface plasmons with light) can be tuned by altering the surface structures.

In Chap. 10, results of field emission studies of ion beam-irradiated metallic thin films were presented. Here the main outcome was a significant enhancement of the field emission current and the field enhancement factor, which was studied as a function of ion beam species and fluence, including the type of metallic substrate such as Al and Cu. For any particular type of substrate material, the enhancement in the field emission current is considered to be due to a combined effect of the micro-conical structures (whose aspect ratio could be varied as a function of ion beam fluence and ionic species) and the embedded subsurface ionic impurities. It would be interesting to carry out future research that would throw further light upon the individual structural and dopant contribution to the electron field emission. It would also be interesting to investigate the effect of heating the substrate while performing the ion beam irradiation experiments, and looking at its effect on the formation of the microstructures. Furthermore, with regard to applications, there is lot of interest in the space plasma physics community on spacecraft charging, where the flux of ions and electrons from the ambient plasma can charge up the spacecraft, leading to (i) damage of onboard electronic instruments and (2) interference with scientific measurements. The first type is very rare but may be harmful. The second type is very common. The energetic fluence of ions over a period of time can lead to development of microstructures on the surface of the spacecraft, which can further exacerbate the surface charging problem and lead to discharge formation. Although Chap. 10 dwells upon this subject, these aspects can be investigated further in greater details, and it would be necessary to work with space relevant materials such as Ti, Al, or their alloys while addressing these issues. The studies on field emission of atomically heterogeneous systems reported in this book can be supported by carrying out multiphysics simulations through software such as COMSOL or using density functional theory. Future work along these directions can throw further light on the role of atomic impurities and micro-conical structures on electron field emission. Environmental surface contamination is another important issue that can degrade the quality of the prepared thin films and affect their intended performance. We have employed residual gas analyzers in our experimental chamber and found that the percentage of carbon is quite small, the main contamination arises from hydrogen and oxygen through small amount of water vapor that may be present inside the chamber. Although the samples are always kept inside the vacuum desiccator, they are needed to be taken out during characterization. A systematic study of the possible sources of introduction of environmental contamination would be very helpful.

References

1. A. Kondyurin, B.K. Gan, M.M.M. Bilek, D.R. McKenzie, K. Mizuno, R. Wuhrerb, Argon plasma immersion ion implantation of polystyrene films. Nucl. Instrum. Methods Phys. Res. B **266**, 1074 (2008)
2. N.L. Singh, Effect of ion beam irradiation on metal particle doped polymer composites. Bull. Matter. Sci. **34**, 81 (2011)
3. I. Lazareva, Y. Koval, M. Alam, S. Stromsdorfer, P. Muller, Graphitization of polymer surfaces by low energy ion irradiation. Appl. Phys. Lett. **90**, 262108 (2007)
4. M.M. Ugeda, I. Brihuega, F. Hiebel, P. Mallet, J.-Y. Veuillen, J.M. Gomez-Rodriguez, F. Yndurain, Electronic and structural characterization of divacancies in irradiated graphene. Phys. Rev. B **85**, 121402 (2012)

5. F. Giannazzo, S. Sonde, V. Raineri, E. Rimini, Irradiation damage in graphene on SiO_2 probed by local mobility measurements. Appl. Phys. Lett. **95**, 263109 (2009)
6. J. Chen, C. Jang, S. Xiao, M. Ishigami, M.S. Fuhrer, Intrinsic and extrinsic performance limits of graphene devices on SiO_2. Nat. Nanotechnol. **3**, 206 (2008)
7. J.-H. Chen, W.G. Cullen, C. Jang, M.S. Fuhrer, E.D. Williams, Defect scattering in graphene. Phys. Rev. Lett. **102**, 236805 (2009)
8. S.H. Al-Harthi, A. Kara'a, T. Hysen, M. Elzain, A.T. Al-Hinai, M.T.Z. Myint, Evolution of surface morphology and electronic structure of few layer graphene after low energy Ar^+ ion irradiation. Appl. Phys. Lett. **101**, 213107 (2012)
9. N. Gorjizadeh, A.A. Farajian, K. Esfarjani, Y. Kawazoe, Spin and band-gap engineering in doped graphene nanoribbons. Phys. Rev. B **78**, 155427 (2008)
10. S. Zhao, J. Xue, Tuning the band gap of bilayer graphene by ion implantation: insights from computational studies. Phys. Rev. B **86**, 165428 (2012)
11. B. Panigrahy, M. Aslam, D. Bahadur, Controlled optical and magnetic properties of ZnO nanorods by Ar ion irradiation. Appl. Phys. Lett. **98**, 183109 (2011)
12. M.K. Gupta, N. Sinha, B. Kumar, dielectric studies and band gap tuning of ferroelectric Cr-doped ZnO nanorods. J. Appl. Phys. **112**, 014303 (2012)
13. K.R. Kittilstved, N.S. Norberg, D.R. Gamelin, Chemical manipulation of high Tc ferromagnetism in ZnO diluted magnetic semiconductors. Phys. Rev. Lett. **94**, 147209 (2005)
14. J.B. Wang, G.J. Huang, X.L. Zhong, L.Z. Sun, Y.C. Zhou, E.H. Liu, Raman scattering and high temperature ferromagnetism of Mn-doped ZnO nanoparticles. Appl. Phys. Lett. **88**, 252502 (2006)
15. O. Mounkachi, A. Benyoussef, A.E. Kenz, E.H. Saidi, E.K. Hlil, High-Tc ferromagnetism in p-type ZnO diluted magnetic semiconductors. Physica A **388**, 3433 (2009)
16. Y. Guo, X.B. Cao, X.M. Lan, C. Zhao, X.D. Xue, Y.Y. Song, Solution based doping of manganese into colliodal ZnO nanorods. J. Phys. Chem. C **112**, 8832 (2008)
17. Z. Zheng, L. Liao, B. Yan, J.X. Zhang, H. Hong, Z.X. Shen, T. Yu, Enhanced field emission from argon plasma-treated ultra-sharp alpha-Fe_2O_3 nanoflakes. Nanoscale Res. Lett. **4**, 1115 (2009)
18. F. Xu, K. Yu, Q. Wang, M. Shi, Q. Zhang, W. Bai, Q. Li, Z. Zhu, Enhanced field emission from H_2 plasma-treated alpha-Fe_2O_3 nanowires. J. Nanosci. Nanotechnol. **7**, 2774 (2007)
19. B.B. Wang, Q.J. Cheng, X.X. Zhong, Y.Q. Wang, Y.A. Chen, K. Ostrikov, Enhanced electron field emission from plasma-nitrogenated carbon nanotips. J. Appl. Phys. **111**, 044317 (2012)
20. J.J.S. Acuna, M. Escobar, S.N. Goyanes, R.J. Candal, A.R. Zanatta, F. Alvarez, Effect of O^{2+}, H^{2+}+O^{2+}, and N^{2+}+O^{2+} ion-beam irradiation on the field emission properties of carbon nanotubes. J. Appl. Phys. **109**, 114317 (2011)
21. J.J.S. Acuna, C.A. Figueroa, D. Biggemann, M.U. Kleinke, F. Alvarez, Oxygen etching mechanism in carbon-nitrogen (CN_x) domelike nanostructures. J. Appl. Phys. **103**, 124907 (2008)
22. M. Kumar, T. Jangid, V. Panchal, P. Kumar, A. Pathak, Effect of grazing angle cross-ion irradiation on Ag thin films. Nanoscale Res. Lett. **11**, 454 (2016)
23. T. Basu, D.P. Datta, T. Som, Transition from ripples to faceted structures under low-energy argon ion bombardment of silicon: understanding the role of shadowing and sputtering. Nanoscale Res. Lett. **8**, 1–8 (2013)
24. J. Majumdar, S. Bhattacharjee, Comparative study on atomically heterogeneous surface with conical arrays of field emitters generated using plasma based low-energy ion beams. J. Appl. Phys. **132** (2022)
25. N. Singh, S. Shah, A. Qureshi, A. Tripathi, F. Singh, D. Avasthi, P. Raole, Effect of ion beam irradiation on metal particle doped polymer composites. Bull. Mater. Sci. **34**, 88 (2011)
26. A. Kondyurin, B. Gan, M. Bilek, D. McKenzie, K. Mizuno, R. Wuhrer, Argon plasma immersion ion implantation of polystyrene films. Nucl. Instrum. Methods Phys. Res., Sect. B **266**, 1074–1084 (2008)
27. I. Chambouleyron, J. Martínez, Optical properties of dielectric and semiconductor thin films, in *Handbook of Thin Films*, (Elsevier, 2002), pp. 593–622

28. Y. Higuchi, H. Tamura, Recent progress on the dielectric properties of dielectric resonator materials with their applications from microwave to optical frequencies. J. Eur. Ceram. Soc. **23**, 2683–2688 (2003)
29. D.E. Aspnes, Optical properties of thin films. Thin Solid Films **89**, 249–262 (1982)
30. N. Sultanova, S. Kasarova, I. Nikolov, Characterization of optical properties of optical polymers. Opt. Quant. Electron. **45**, 221–232 (2013)
31. M. Y. Koledintseva, R. E. DuBroff, and R. W. Schwartz, Maxwell Garnett model for dielectric mixtures containing conducting particles at optical frequencies, Technical report, Progress in Electro magnetics Research (2006)
32. J. Pitarke, V. Silkin, E. Chulkov, P. Echenique, Theory of surface plasmons and surface-plasmon polaritons. Rep. Prog. Phys. **70**, 1 (2006)
33. E. Economou, Surface plasmons in thin films. Phys. Rev. **182**, 539 (1969)
34. W.L. Barnes, A. Dereux, T.W. Ebbesen, Surface plasmon subwavelength optics. Nature **424** (2003)
35. J. Homola, S.S. Yee, G. Gauglitz, Surface plasmon resonance sensors. Sensors Actuators B Chem. **54**, 3–15 (1999)
36. A. Degiron, T. Ebbesen, The role of localized surface plasmon modes in the enhanced transmission of periodic subwave length apertures. J. Opt. A Pure Appl. Opt. **7**, S90 (2005)
37. J. Zhang, L. Zhang, W. Xu, Surface plasmon polaritons: physics and applications. J. Phys. D. Appl. Phys. **45**, 113001 (2012)

The manufacturer's authorised representative in the EU is Springer Nature Customer Service Centre GmbH, Europaplatz 3, 69115 Heidelberg, Germany. If you have any concerns regarding our products, please contact ProductSafety@springernature.com

Printed and bound by CPI Group (UK) Ltd, Croydon, CR0 4YY
21/09/2026
02219978-0005